D1241927

INTRODUCTION TO BIOCATALYSIS USING ENZYMES AND MICRO-ORGANISMS

This is an introductory text intended to give the non-specialist a comprehensive insight into the science of biotransformations. The book traces the history of biotransformations, clearly spells out the pros and cons of conducting enzyme-mediated versus whole-cell bioconversions, and gives a variety of examples wherein the bio-reaction is a key element in a reaction sequence leading from the cheap starting materials to valuable end-products (such as pharmaceuticals and agrochemicals, fragrances and flavours). Biotransformations involving the hydrolysis of esters, amides, and nitriles, the synthesis of esters and amides, reduction and oxidation reactions, and carbon-carbon bond-forming systems are discussed. The book finishes with a discussion of some industrially important large-scale bioconversions.

INTRODUCTION TO BIOCATALYSIS USING ENZYMES AND MICRO-ORGANISMS

STANLEY M. ROBERTS
University of Exeter

NICHOLAS J. TURNER
University of Exeter

ANDREW J. WILLETTS
University of Exeter

MICHAEL K. TURNER
University College London

CAMBRIDGE
UNIVERSITY PRESS

Published by the Press Syndicate of the University of Cambridge
The Pitt Building, Trumpington Street, Cambridge CB2 1RP
40 West 20th Street, New York, NY 10011-4211, USA
10 Stamford Road, Oakleigh, Melbourne 3166, Australia

First Published 1995

Printed in the United States of America

Library of Congress Cataloging-in-Publication Data

Introduction to biocatalysis using enzymes and micro-organisms / S.M.
Roberts ... [et al].
p. cm.
Includes bibliographical references and index.
ISBN 0-521-43070-4. – ISBN 0-521-43685-0 (pbk.)
1. Microbial biotechnology. 2. Enzymes – Biotechnology.
3. Biotransformation (Metabolism) I. Roberts, Stanley M.
TP248.27.M53157   1995
660'.63 – dc20   93-48247
                                                 CIP

A catalog record for this book is available from the British Library.

ISBN 0-521-43070-4 Hardback
ISBN 0-521-43685-0 Paperback

Contents

Preface

Twenty years ago, only a handful of laboratories, some in industry and some in academia, were interested in using enzyme-catalysed reactions (biotransformations) in organic synthesis. At best those researchers were considered to be at the margin of mainstream synthetic organic chemistry; at worst they were considered to be downright odd.

In the 1980s there was an exponential increase in interest in the area of biotransformations. That worldwide increase in attention to this field of research can be assigned to several factors, including the perception and perseverance of the original researchers in the field, the increased availability of a wide variety of enzymes, and the realization that many families of enzymes will transform a wide range of unnatural compounds, as well as their natural substrates.

The utilization of enzymes in organic synthesis can be advantageous for several reasons:

1. Enzymes catalyse reactions under mild conditions with regard to temperature (ca. 37 °C), pressure (1 atm), and pH (ca. 7.0). The transformations are often remarkably energy-efficient when compared with the corresponding chemical processes.
2. Enzymes often promote highly chemoselective, regioselective, and stereoselective reactions, and being chiral catalysts, they are often able to generate optically active compounds. The increased awareness regarding the need to have optically pure compounds for such uses as pharmaceuticals and agrichemicals (so as to avoid unnecessary toxicity and/or ecological damage) has been a significant driving force in the development and exploitation of non-natural enzyme-catalysed reactions.
3. Enzymes can promote reactions that are difficult or impossible to

emulate using other techniques of synthetic organic chemistry. Thus bioconversions can generate new series of chiral synthons and, in other cases, may allow a short cut to be taken in a known synthetic sequence.

4. Enzymes are natural catalysts, and this can be advantageous where "green appeal" is a commercial benefit and/or an ecological requirement.

5. Enzymes use water as the reaction medium, and as the disposal of organic solvents becomes progressively more difficult, this may become increasingly important. On the other hand, some enzymes (e.g. lipases) can function as catalysts in organic media, thus allowing a choice to be made regarding the preferred solvent system.

With the massive increase in interest in the field of biotransformations, many scientists are being attracted to the area. Moreover, the subject is being introduced into courses at higher-education establishments, at both the graduate and post-graduate levels.

With these recent developments in mind, it seems timely to introduce a text covering the basic groundwork of the subject. The first chapter of this book sets out the fascinating history of biotransformations. Because it is the first time that this story has been put together, comprehensive references are provided for Chapter 1. The second chapter describes the theory and experimental practice of enzyme-catalysed microbial biotransformations. This is the basis of the science of biotransformations, and considerable detail is given. Thereafter, Chapters 3–5 cover the main areas of interest in the field of biotransformations: Not only are the biotransformations themselves described, but also the importance of the derived products in the synthesis of fine chemicals is illustrated. The examples chosen are not particularly special in themselves; there are many similar cases in the literature that we have chosen not to mention. Thus our text is certainly not comprehensive, and so references to sources and reviews are given for Chapters 3–5 to help the reader gain further familiarity with the subject.

Earlier, one of the perceived difficulties in the area of biotransformations was that of scale-up. Thus, while it seemed that test-tube experiments probably would be entirely feasible, the use of enzyme-catalysed processes on a kilogram scale or a tonne scale seemed a daunting task. It seems that, in many cases, those fears were unfounded, and the final chapter deals with some of the large-scale processes that have been industrial successes in terms of engineering and commerce. Because previously this topic has not been reviewed as extensively as the material described in Chapters 3–5, more detailed descriptions of the processes have been included. The latter part of the final chapter describes some special techniques of increasing importance in the field of biotransformations.

It seems that enzyme-catalysed reactions and whole-cell-mediated transformations are set to make a major impact in synthetic organic chemistry over the next few years. On the more distant horizons, the complementary field of catalytic antibodies is beginning to demonstrate its potential power but that is another story which may well mushroom so as to necessitate the writing of other specialist text-books devoted to that topic.

There are many people, too numerous to name, who have contributed, through discussion, to the writing of this book. We thank them all, but we would like to single out Mrs. Claire Turner of the Museum of Brewing at Burton-on-Trent, Dr. Dennis Briggs of the School of Biochemistry at the University of Birmingham, and Dr. John Woodley of the Department of Chemical and Biochemical Engineering at University College London for their help and advice with Chapter 1. We would also like to acknowledge the assistance of the Royal Society of Chemistry and the Institute of Brewing for allowing us free access to their library archives.

Bibliography

General background reading

Biotransformations in Organic Chemistry: Principles and Applications.
    Chimia, 47 (1993), 5–68.
Faber, K. (1992). *Biotransformations in Organic Chemistry.* Springer-Verlag, Berlin.
Poppe, L., & Novak, L. (1992). *Selective Biocatalysis: A Synthetic Approach.* VCH
    Publishers, Weinheim.
Servi, S. (1992). *Microbial Reagents in Organic Synthesis.* Kluwer Academic,
    Dordrecht.
Kang, A. S., Kingsbury, G. A., Blackburn, G. M., & Burton, D. R. (1990). Catalytic
    Antibodies – Designer Enzymes. *Chemistry in Britain*, 128.
Abramowicz, D. A. (1990). *Biocatalysis.* Van Nostrand Reinhold, New York.
Suckling, C. J. (1990). *Enzyme Chemistry: Impact and Applications*, 2nd ed.
    Chapman & Hall, London.
Lehninger, A. L., Nelson, D. L., & Cox, M. M. (1993). *Principles of Biochemistry.*
    Worth, New York.

Description of a wide range of validated experimental procedures

Roberts, S. M., Wiggins, K., & Casy, G. (1992). *Preparative Biotransformations:
    Whole Cell and Isolated Enzymes in Organic Synthesis.* Wiley, Chichester.

Abbreviations

ADP	adenosine diphosphate
ATP	adenosine triphosphate
CMP	cytidine monophosphate
CM-cellulose	carboxymethylcellulose
CTP	cytidine triphosphate
DFP	diisopropyl phosphorofluoridate
DHAP	dihydroxyacetone phosphate
DMF	dimethylformamide
DNA	deoxyribonucleic acid
EDTA	ethylenediaminetetraacetate
EGTA	(ethylenedioxy) diethylene nitrilotetra acetate
HIV	human immunodeficiency virus
HLAD	horse liver alcohol dehydrogenase
HSAD	$3\alpha,20\beta$-hydroxysteroid alcohol dehydrogenase
NAD$^+$	nicotinamide-adenine dinucleotide
NADP$^+$	nicotinamide-adenine dinucleotide phosphate
PEG	polyethylene glycol
PHB	polyhydroxybutyrate
RAMA	rabbit muscle aldolase
TBAD	*Thermoanaerobium brockii* alcohol dehydrogenase
TEPP	tetraethyl pyrophosphate
UDP	uridine diphosphate
YAD	yeast alcohol dehydrogenase

1

An historical introduction to biocatalysis using enzymes and micro-organisms

1.1 Background

Each year that passes sees increasing interest in the application of enzymes and micro-organisms as catalysts in organic chemistry. Well-received national reports on the potential growth of biotechnology, not to speak of numerous reviews of the topic of biotransformation itself, have described the advantages which biological catalysis can bring to complex organic syntheses. At this time it is indeed an exciting prospect for anyone who has a rôle to play in the development of this new technology. But occasionally a prospective participant, as well as a reviewer, ought to stand back and ask, Why has this explosion of interest happened now?

An answer to such a question often lies in the historical context from which technologies and their underlying science develop. This is a topic worthy of greater attention than it usually receives from practising scientists. A resort to scientific papers more than one or two decades old is frequently viewed as perverse, while to consult the scientific literature of the past century in search of a pertinent contemporary lesson is simply eccentric. Yet this old literature can be worth reading for the light it can shed on current research, quite apart from which, it is, more often than not, a pleasure to read.

The biochemistry of the late nineteenth century has an interesting connection with the manufacture of vitamin C. The basis of this process relied on the development of the concepts of fermentation and catalysis, but it emerged directly from a microbial oxidation described in a short paper written by Adrian Brown (1886a). The background to his work, and its industrial setting, can provide the focus for a broader introduction to the use of enzymatic catalysis in modern organic chemistry. It will also suggest an answer to the question posed earlier.

For anyone interested in reading more extensively, a wide selection of the relevant nineteenth-century literature is not usually available, except in the older universities. However, the books by Boyde (1980) and by Teich and Needham (1992) contain many of the key papers translated into English, connected with a commentary.

1.2 Kirchhoff and starch hydrolysis

In the early nineteenth century, developments in biochemistry were closely associated with those in organic chemistry. Moreover, the science of these subjects was not distinct from the technology of their application. The chemical and biochemical reactions in which starch and sugar participated were important topics for scientific research and technological development. The trail from that period to the current methods for manufacture of ascorbic acid (vitamin C) may as well begin with a note about the ink with which a chapter such as this would once have been written. It would at one time have contained gum arabic, which was, and still is, an expensive natural carbohydrate. As an alternative, Bouillon-Lagrange (1811) described a method for preparing inks using a modified starch. He heated dry starch powder gently until it began to char. This material formed a suitable gum in water, which he noted to have a sweet taste. He further treated the gum with acids, but they were too strong to have enhanced the sweet flavour, since sulphuric acid caused further charring and gave off a smell of acetic acid; in contrast, nitric acid formed oxalic acid.

The preparation of that gum is worth recording, because in the following year Kirchhoff showed that if starch was boiled in dilute sulphuric acid, the suspension became sugary (Teich and Needham, 1992). That sugar (Vogel, 1812) was soon shown to be fermentable; moreover, the same acid treatment converted the sugar in milk [lactose (1)] into a fermentable form through its hydrolysis into galactose (2) and glucose (3) (Scheme 1.1) (Vogel, 1817).

It is easy to write in terms of definite chemicals entities after an interval of nearly two centuries. It is wrong, even demeaning of their achievements, to imagine those chemists as having anything more than their contemporary understanding of the experiments. Nevertheless, the importance of this discovery, that the acid treatment of starch would convert it to sugar, was clearly recognized. Not only was the conversion

interesting in its own right, perhaps in shedding light on the chemistry and physiology of several processes in plants; but it also gives society a product, which

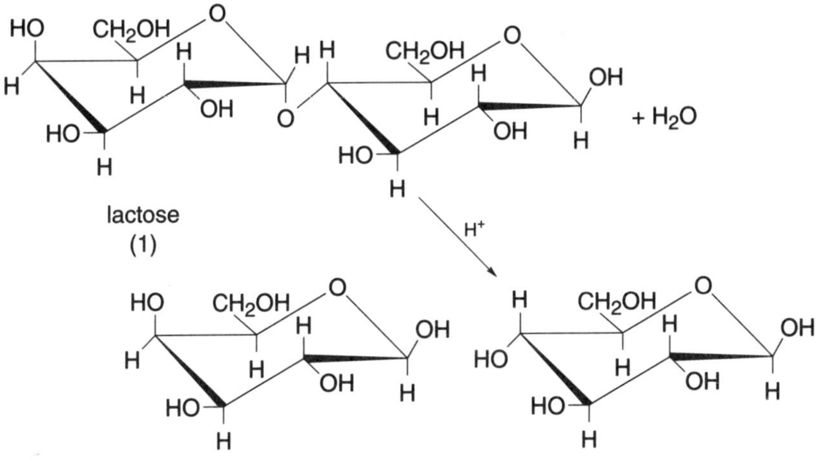

Scheme 1.1. Hydrolysis of lactose.

in many circumstances could replace the cane sugar, which is already expensive, and whose price rises incessantly.
(De Saussure, 1814, p. 499)

At the time, organic chemists found these materials very difficult to study. The simple elemental ratios which were the rule in the compounds of inorganic origin were not found in organic chemistry. Nevertheless, it was clear that in decomposing the starch into sugar, the sulphuric acid did not enter into the product, nor was it consumed in the process. It was correctly inferred from the elemental analysis that only the elements of water were entering the starch during the conversion (Table 1.1).

The influence of the acid in this operation appears to be limited to increasing the fluidity of the aqueous solution of the starch, so helping the latter to combine with water.
(De Saussure, 1814, p. 501)

The fractional ratios between the elements in the materials isolated from living matter convinced some that the chemical processes associated with their synthesis were somehow different from those associated with inorganic matter and incorporated some vital factor:

All simple bodies in nature are subject to the action of two powers, of which one, that of attraction, tends to unite the molecules of bodies one with another, while the other, produced by caloric, forces them apart.... A certain number of these

Table 1.1. *Elemental analysis of starch and of the sugar recovered after treatment with sulphuric acid*

Sample	C	O	H	N
Starch				
De Saussure observed	45.39	48.31	5.90	0.40
$(C_6H_{10}O_5)_n$ theory[a]	44.5	49.3	6.2	—
Released sugar				
De Saussure observed	37.29	55.87	6.84	—
$C_6H_{12}O_6$ theory[a]	40.0	53.3	6.7	—
$C_6H_{12}O_6 \cdot \frac{1}{2}H_2O$ theory[a]	38.1	55.0	6.9	

[a]Calculated values based on the given elemental compositions.
Source: De Saussure (1814).

simple bodies in nature are subject to a third force, to that caused by the vital factor (*le principe vital*), which changes, modifies and surpasses the two others, and whose limits are not yet understood.
(*Beral, 1815, pp. 358–9*)

   In the discussion which followed that paper (p. 361), one of the journal's editors, J.-J. Virey, pointed out that there were some products of inorganic chemistry that also had poorly defined compositions, notably the oils formed when cast iron was treated with acid, or when olefinic gases were burnt. He also pointed out that some chemical processes (e.g. Kirchhoff's conversion of starch into sugar with acid) resembled the effects of germination and that respiration could release simple materials such as carbonic acid and water. Thus the distinction between organic and inorganic chemistry was not as clearly defined as Beral had portrayed it, and it warranted further study.

   The chemical literature throughout the nineteenth century uses the term "body" in a sense which is similar to our contemporary use of the term "compound". (More generally, it is not prudent to assume that our modern definitions of chemical terms are the same as those once current.) Virey's remarks clearly illustrate one other feature of the chemistry of the early nineteenth century: that its experiments were much affected by impure reagents, even though the inferences drawn from them ring true in the late twentieth century.

1.3 Payen, Persoz and diastase

This early history of the hydrolysis of starch is the chemical background against which the technical study of catalysis and enzymology began.

Within a few years the conversions of starch and lactose were repeated with biological agents. Kirchhoff (1816) described the saccharification of starch in grain, and Vogel (1817) showed that an infusion of oats would produce a fermentable sugar from milk. Döbreiner (1815) used a wet yeast paste to convert sugar into "une liquer vineuse" which was no longer sweet.

The brewing industry must have played some part in determining the content of these experiments. It was already a large industry; Samuel Whitbread brewed 30,000 m^3 of beer in 1796. Moreover, the manufacture of ethanol was developing as a separate industry alongside brewing. In 1830 Aeneas Coffey, in Dublin, developed a distillation process with a multiplate countercurrent condenser (Figure 1.1) capable of recovering ethanol as an azeotrope with water from a fermented beer mash (Packowski, 1978). Biotechnology may have an ancient history, but this process, which incorporates the first large-scale use of a biochemical step in the manufacture of an organic reagent, can properly lay claim to be the first biotechnological process of modern chemistry.

The successful development of this technology is remarkable in the context of the very limited understanding of the process itself. The descriptions of nineteenth-century brewing practice suggest that it was anything but reliable; at Burton-on-Trent in Britain, little beer was brewed in the summer months (Brown, 1916) because the chance of spoilage was so high. The unreliable nature of the biocatalytic process was a direct consequence of this ignorance.

The first advance in understanding the process occurred in 1833 when Payen and Persoz extracted a mixture of amylases from malted barley. This soluble material ("soluble ferment") was able to separate and saccharify the starch away from the husk of the grain. They called the activity "diastase" (from the Greek word meaning "to separate"). The work was significant because it demonstrated that the "soluble ferment" could be dissolved out of the organised structures of the malted barley or the yeast, the so-called organized ferments. The saccharification of the starch which Kirchhoff had described could then be studied separately from the fermentation itself. The term "diastase" was eventually to be used as a description of any soluble enzyme, although it later reverted to an alternative description of an amylase.

Their research led to technological improvements in brewing practice, in which those authors participated:

It will seem less remarkable that we have advanced only such a little distance along this new road if it is considered that we have been caught up in a millrace of

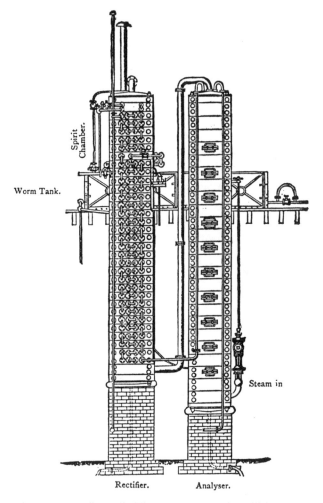

Figure 1.1. Coffey's distilling apparatus. The still has two columns, the analyser and the rectifier. Both are multiplate distillation units. Steam enters at the base of the analyser (Steam in), through which it rises. From the top it is piped to the lowest compartment of the rectifier, through which it again rises. The alcoholic mash enters through the top of the rectifier, through which it falls inside a long zigzag pipe which acts as a heat exchanger. From the bottom of the rectifier it is piped to the top of the analyser, through which it falls freely, mixing with the rising steam. It eventually leaves from the base of the analyser. The effect of this countercurrent process is to strip alcohol from the mash falling through the analyser against the flow of the steam. The multiplate arrangement ensures efficient removal of the alcohol from the mash. The alcohol-laden steam leaves the top of the analyser and enters the base of the rectifier, through which it rises against the flow in the zigzag cooling pipe containing the fresh mash. The multiple arrangement inside the rectifier ensures many stages of boiling and condensation so that the vapour at the top of the column is an azeotrope of water and ethanol. The upper section of the rectifier is the spirit

newborn applications, and that we have not thought it right to refuse our collaboration to the manufacturers who have requested it from every direction.
(*Payen and Persoz, 1833, p. 92*; see Boyde, 1980)

This was not surprising. Within two years there was an extensive paper comparing the effects of diastase and acid on a variety of starches, leading to the preparation of crystalline sugar (Guérin-Varry, 1835).

1.4   Berzelius and catalysis

The importance of the work by Payen and Persoz was clear to Berzelius. When he thoroughly revised his *Textbook of Chemistry* for the fourth German edition (Berzelius, 1838; the previous edition is dated 1831), he included the famous section on catalysis which he originally put forward in 1828 (Jorpes, 1966). Chemists were clearly puzzled by the organic processes in the living tissues. Berzelius compared them to a group of decomposition phenomena (Table 1.2) which could not be explained by the process of "double decomposition" in which two compounds reacted together. He suggested that some substances were characterized by

a new power to produce chemical activity belonging to both inorganic and organic nature [which] using a derivation well-known in chemistry [he called] the catalytic power of the substances, and decomposition by means of this power catalysis, just as we use the word analysis to denote the separation of the component parts of bodies by means of ordinary chemical forces.
(*Berzelius, 1838; see Jorpes, 1966*)

The implication that catalysis was a degradative power is intriguing, just as is the old use of the term "analysis," instead of the modern word "synthesis". It is also interesting to see what now we would regard as chemical and biochemical examples of catalysis freely taken together. The action of fibrin in degrading hydrogen peroxide (Table 1.2) presumably was due to a small amount of catalase with which it was contaminated.

Elsewhere (Jorpes, 1966), Berzelius wrote of diastase that:

One can hardly assume that this catalytic process is the only one in the vegetable kingdom. On the contrary, it gives reason to believe that within living plants and

◄ ───

chamber, in which the constantly boiling mixture of the azeotrope is refluxed. The azeotrope leaves the rectifier from the bottom of this refluxing section and runs to the worm tank. The feints (a dilute solution of ethanol in water) which collect at the base of the rectifier are piped to the top of the analyser, there to be mixed with the mash. (From Nettleton, 1893; reproduced with the kind permission of the Institute of Brewing.)

Table 1.2. *Catalytic processes listed by Berzelius in 1838*

Catalytic process	Source of research
Acidic hydrolysis of starch to sugar	Kirchhoff
Decomposition of hydrogen peroxide in acid and alkali	Thenard
Decomposition of hydrogen peroxide by platinum and manganese dioxide	
Decomposition of hydrogen peroxide by fibrin	
Combustion in air of alcoholic and ethereal vapours on platinum	H. Davy
Effects of finely divided platinum	E. Davy
Ignition of hydrogen in air on platinum sponge	Döbreiner
Conversion of sugar into carbonic acid during yeast fermentation	

animals thousands of catalytic processes are going on between the tissues and the fluids, producing a multitude of chemical compounds, the creation of which out of the common raw material, the sap of plants or blood, has up to now been unexplained, and which may possibly be found in the future to depend on the catalytic power of the living tissues.
(*Berzelius, 1838; see Jorpes, 1966*)

1.5 Leibig and Hofmann

Berzelius had a poor view of British chemists (Jorpes, 1966). They nevertheless understood the importance of what he had written. When 10 years later Playfair (1848) reviewed the topic, he also took the decomposition of sugar by yeast as one of the examples with which to illustrate his views on the mechanism of catalysis. Perhaps this interest was partly the result of an influx of German chemists. In 1845, at Leibig's recommendation, Hofmann was appointed the director of the new Royal College of Chemistry in London (Travis, 1992; Leaback, 1992). That same year, another of Leibig's students, Henry Böttinger, left Germany, eventually to join the brewers Allsopp & Sons at Burton-on-Trent, where he was quickly appointed the head brewer.

Böttinger became friendly with Hofmann, and the students at the Royal College of Chemistry benefitted from the quantities of beer which he sent for analysis. When Payen suggested that the British brewers added strychnine to increase the bitterness of their beer, Böttinger turned to Hofmann for help. The report was written jointly with Thomas Graham, who was the professor of chemistry at University College London (Hofmann and Graham, 1852). It is, in retrospect, mildly amusing, because of the absurdity of the claim, which is in contrast to the magisterial tone of its dismissal, but at the time it was no joke. The brewing of beer was a major industry throughout Europe, and the report was widely circulated (Armstrong, 1921, p. 244). Payen maintained that he had been misquoted by a journalist in France, a view which Hofmann and Graham accepted. Obviously the inaccurate reporting of scientific matters in the press is an old problem!

Neither Böttinger nor Hofmann was likely to understand the fermentation process while their outlook was limited by the all-pervasive views of their teacher, Leibig (Brown, 1916; Armstrong, 1921). He believed that it represented an inanimate interaction between the motions of the bodies of the yeast and the organic reagents. In that respect, Leibig's views seem not to be very different from those of Beral (1815) stripped of their reference to caloric.

Hofmann edited the English translation of Leibig's *Annual Report of Progress in Chemistry*. The report for 1847–8 contains details of Pasteur's physical separation of the crystals of the two isomers of tartaric acid, while that for 1849 reports Leibig's experiments on the fermentative conversion of malate into succinate. In an experiment lasting about three months, Dessaignes (1849) noted that crystals of calcium malate, held under water, were slowly converted into succinate. Leibig (Leibig and Kopp, 1849) showed that beer yeast, putrid fibrin and rotten cheese would also catalyse this change. As much as 15 or 16 ounces (about 0.43 kg) of succinic acid were obtained from 3 pounds (about 1.3 kg) of crude malate of lime

Malic acid.   Succinic acid   Acetic acid

$$6\,C_4H_2O_4 + 4\,HO = 4\,C_4H_3O_4 + C_4H_4O_4 + 4\,CO_2$$

Malic acid.   Butyric acid.

$$4\,C_4H_2O_4 + 4\,HO = C_8H_8O_4 \;+\; 8\,CO_2 \;+\; 4H$$

Succinic acid.   Butyric acid.

$$3\,C_4\,H_3\,O_4 \;\;=\;\; C_3H_8O_4 + 4\,CO_2 \;+\; H$$

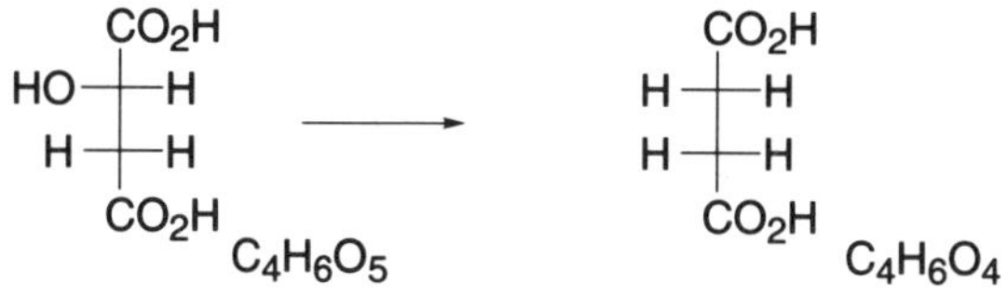

L-(S)-malate   succinate

Scheme 1.2. The upper section of the scheme is taken directly from the original publication.

(calcium malate). Under other conditions, substantial amounts of acetic acid and butyric acid were also formed (Scheme 1.2). There is little reason to doubt these results, despite the problems with the elemental analysis of the products. (It is also noteworthy that the modern organic chemist would be more interested in the introduction of the chiral center than in its removal.)

Such transformations appeared regularly in the literature of the time, often as brief reports. Phipson (1862) compared the chemical and biological oxidations of citric acid. Permanganate was the chemical oxidant which produced oxalic acid; uncooked putrid beef and boiled beef were the biological agents which yielded butyric acid (Scheme 1.3). It is not surprising that the analysis of these transformations (Scheme 3) should differ from a modern interpretation.

Citric acid ($C_6H_8O_6$)

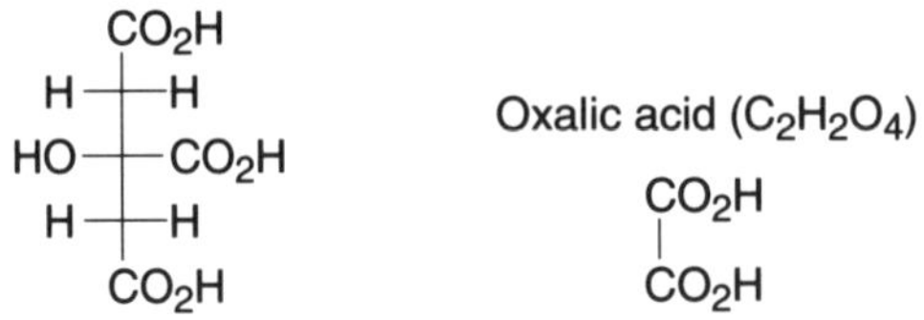

Oxalic acid ($C_2H_2O_4$)

$$C_{12}H_8O_{14} + O_{12} = 3C_4H_2O_8 + 2HO.$$
Citric Acid   Oxalic Acid

$$C_{12}H_8O_{14} = C_8H_8O_4 + 4CO_2 + O_2$$
Citric Acid   Butyric Acid

$$CH_3CH_2CH_2CO_2H$$
Butyric acid ($C_4H_8O_2$)

Scheme 1.3. The equations (only) are taken from the original publication.

The state of mid-nineteenth-century chemistry and its impact as a technology are nicely recorded in a lecture which Hofmann addressed to teachers (Hofmann, 1861). He admitted that the study of organic chemistry was in its infancy, but he foresaw its potential:

The notion that the action of most of our medicines is chemical, is daily growing into a general conviction. We admit that with every change wrought by pharmaceutical agents in the state of our organism, there occurs a corresponding change in its composition, resulting from their reaction on one or more of its constituents. . . . Associated with chemistry, medicine no longer draws the veil of vitality over processes, the mystery of which may be unlocked by the key of analysis. . . .

The special zeal with which the field of organic chemistry has been cultivated during the last thirty years, the simple and accurate methods which we now possess for determining the composition of organic products, the amount of analysis actually performed, and, more than all, the still untiring energy of the numerous labourers in the same field of investigation, hold out the promise that the connexion between medicine and chemistry, becoming daily more intimate, will be productive of benefits, the importance of which we can scarcely venture to estimate in the present state of our knowledge.
(Hofmann, 1861, pp. 12–13)

It is interesting to note the importance which Hofmann assigned to "analysis" and what appears to be a change in its definition compared with that in the quotation from Berzelius.

1.6 Pasteur and organic chemistry

Quite apart from the chemical problems which this work posed, it also suffered from the inadequate contemporary understanding of the biology on which the fermentation technology was based. Fermentations were variable, a factor of considerable economic consequence to the brewers, vintners and distillers alike. This was the problem which Bigo (a local distiller) brought to Pasteur in 1856 shortly after the latter's appointment as professor and dean of the new Faculty of Sciences at Lille in France. Bigo manufactured alcohol from sugar beet, but the fermentations would often become acidic and would yield lactic acid rather than ethanol. It must have been a problem such as this which Fortoul, the French minister for public instruction, had in mind when he wrote to Guilleman, the rector at Lille, that

Pasteur must guard against being carried away by his love for science, and he must not forget that the teaching of the faculties, whilst keeping up with scientific theory, should, in order to produce useful and far-reaching results, appropriate to itself the special applications suitable to the real wants of the surrounding country.
(Vallery-Radot, 1901, Chap. 4)

(Today's politicians would seem to give themselves too much credit for the novelty of their views.)

The work which Pasteur did for Bigo is usually credited with firing his enthusiasm for fermentation. He discovered (Pasteur, 1858b) a new organism (*levûre lactique*) whose cells were much smaller than those of brewer's yeast, and which was always associated with lactic acid fermentation. In another experiment he recovered a micro-organism from aqueous suspensions of crude calcium tartrate which would ferment tartaric acid [D-tartrate (4)]. The process was easy to follow with a polarimeter as the tartrate was decomposed. However, when he repeated the experiment with racemic acid (DL-tartrate), only the D-tartrate was consumed, and he was able to recover crystalline L-tartrate (5) (Pasteur, 1858a). He noted that this was an excellent method of preparing L-tartrate, and two years later, after repeating the process with *Penicillium glaucum*, he wrote that "it recommends itself as a method, probably of very general application, for splitting apart organic bodies in which it would be reasonable to suppose a molecular composition of the same nature as that of paratartaric acid" (Pasteur, 1860, p. 299). In doing this, Pasteur had put in place the third of his methods for separating optical isomers, following the actual separation of the different crystals (object and mirror-image forms) and the use of an optically active natural base such as quinine to form diastereomeric salts.

$$
\begin{array}{cc}
\mathrm{CO_2H} & \mathrm{CO_2H} \\
\mathrm{H}\!-\!\!-\!\mathrm{OH} & \mathrm{HO}\!-\!\!-\!\mathrm{H} \\
\mathrm{HO}\!-\!\!-\!\mathrm{H} & \mathrm{H}\!-\!\!-\!\mathrm{OH} \\
\mathrm{CO_2H} & \mathrm{CO_2H} \\
\text{L-(+)-(}R,R\text{)-} & \text{D-(–)-(}S,S\text{)-} \\
\text{(4)} & \text{(5)}
\end{array}
$$

tartaric acid

It seems likely that Pasteur's original reason for performing some of these microbiological experiments had more to do with his study of the effects of optically active nutrients on the forms of the organisms which could use them, rather than as an investigation of the fermentation itself (Root-Bernstein, 1989). He wondered whether molecular form could influence the shape of living organisms in the same way as it did the shape of crystals. The outcome of the experiments that followed was rather different, but no less significant. He realized that just as the nature of the yeast, either brewer's or lactic, would divert the fermentation of sugar to alcohol or to lactic acid, so the nature of the nutrient on which the yeast grew could also affect the course of the fermentation. Observations such as these were to lead to his proposition that the course of each fermentation was determined by the organisms which it contained, and that the process itself was dependent on the action of a living organism.

It is necessary to ignore most of what is now taken for granted about fermentation to imagine the powerful effect of Pasteur's work. The advance in the latter half of the twentieth century which has had an equivalent effect on the science and technology of microbiology must surely be the discovery of the connection between DNA and the genetic code. They brought about similar revolutions in microbiology. Pasteur's concept of the fermentation process itself was in complete contrast to that of Leibig, whose mechanistic attitude to the action of yeast actually delayed advances in brewing practice:

... the a priori ideas of Leibig ... owing to the genius and commanding authority of their great apostle, had acted as a bar to progress and prevented a dispassionate consideration of the new vitalistic theories of fermentation which were destined to revolutionise all our conceptions of such phenomena.
(Brown, 1916, p. 275)

There is no need here to describe the dispute on the nature of fermentation which subsequently arose between Leibig and Pasteur. There are interesting accounts of the influence of the debate reported by their contemporaries (Sykes, 1895; Frankland, 1897; Brown, 1916). It was finally resolved 20 years later with the publication of *Etudes sur la bière* (Pasteur, 1876), which set out a modern theory of fermentation practice.

The influence of Pasteur's work on organic chemistry is underrated compared with his influence on the study of fermentation and infectious disease. Fermentation became a source of compounds for study, and the separation of isomers was extended to a much larger range of compounds and reactions. Plimpton (1881) prepared a series of amines from chiral and achiral pentanol (amyl alcohol), and Frankland (1885) was able to review a number of chemical changes in their relation to micro-organisms. He noted that there were two types of chemical changes, those effected when two or more substances came into contact, and those effected by contact with a substance which itself was unaltered:

Failing any satisfactory explanation, very heterogeneous changes of the latter kind have been grouped together under the designation of "catalytic reactions", but a careful study of many of the reactions of this second class has transferred them to the first, and it is more than probable that the remainder, when better known, will be similarly disposed of. The chemical changes occurring in animal and vegetable organisms were, until recently, tacitly, if not formally relegated to the second type. The plant or animal was regarded as effecting the changes by mere contact, or by some mysterious process outside the ken of experimental enquiry. This illusion has been finally dispelled by the synthetical operations of organic chemistry, which have taught us how to produce, by purely laboratory processes, numerous compounds formerly obtainable only as the products of living organisms, and it is to be hoped that chemists and biologists will now give more attention

Table 1.3. *Frankland's list of soluble ferments (1885)*

Soluble ferment	Substrate	Products	Enzyme present in soluble ferment
Diastase	Starch	Dextrin & glucose	Amylase
Invertin	Cane sugar	Glucose & laevulose	Invertase
Synaptase	Salicin	Glucose & saligenin	Glucosidase
Emulsin	Amygdalin	Glucose, benzoic acid & HCN	
	Arbutin	Glucose & hydroquinone	
	Helicin	Glucose & salicylic hydride	
	Phloridzin	Glucose & phloretin	
	Esculin	Glucose & esculetin	
	Daphnin	Glucose & daphnetin	
Pancreatic ferment	Fat	Margaric acid & glycerin	Lipase

to this hitherto neglected field of chemical action – the chemical changes which occur in animal and vegetable organisms.
(Frankland, 1885, p. 159)

Frankland believed that the chemical actions of living organisms were of two kinds, synthetical and analytical (in the sense used by Berzelius), the first being chiefly performed by plants, and the latter by animals, with micro-organisms belonging to the second category. He reviewed a range of processes effected on the one hand by the soluble ferments (Table 1.3), and on the other by the micro-organisms themselves. He noted that all of these processes had their chemical analogies, but that the "organised ferments" were able to carry out processes for which there was no direct chemical equivalent, although there might be some indirect chemical method of performing the same transformation. Most of the reactions which he listed in the latter category were fermentations producing succinate or butyrate from a range of substrates, amongst them Leibig's experiment already described and the synthesis of succinate and glycerol from glucose (Scheme 1.4). By now the atomic composition was correct even if the molecular structure was not appreciated. However, he specifically excluded from this catalogue the power of micro-organisms "to destroy one of the optically active compounds in a mixture, and thus to isolate the compound of opposite optical activity" (Frankland, 1885, p. 181).

$$3 \begin{cases} COHo \\ C_2H_3Ho \\ COHo \end{cases} = 2 \begin{cases} COHo \\ C_2H_4 \\ COHo \end{cases} + \begin{cases} CH_3 \\ COHo \end{cases} + 2CO_2 + OH_2.$$

Malic acid   Succinic acid Acetic acid

$$4C_6H_{12}O_6 + 3OH_2 = \begin{cases} COHo \\ C_2H_4 \\ COHo \end{cases} + \begin{cases} CH_2Ho \\ CHHo \\ CH_2Ho \end{cases} + 2CO_2 + O.$$

Glucose   Succinic acid Glycerin

Scheme 1.4. The equations are reproduced directly from the original publication.

Reviewing Pasteur's separation of racemic acid, he noted that "Le Bel afterwards extended these observations to optically active amyl alcohol and propyl glycol, whilst Lewkowitsch was the first who showed that by varying the microorganisms, either the dextro- or laevo-compound (in the case of mandelic acid) could be destroyed at will" (Frankland, 1885, p. 181).

Clearly the study of the chemical transformations which micro-organisms and enzymes would catalyse was a fitting study for organic chemists. Nevertheless, they were not entirely familiar with the good microbiological practice necessary for reproducible experiments, and a physician from St. Bartholomew's Hospital reminded them that

not till chemists come to look on the matter in the same light in which we look upon it, namely to obtain the organisms pure, to render nutritive material sterile, to be able to produce with this pure organism the specific chemical activity you wish to obtain – not till you have fulfilled all these conditions, can you claim to have established the fact that a definite organism is the cause of a definite chemical process.
(Klein, 1886, p. 205)

1.7 Burton-on-Trent and brewing

These developments in microbiology were not lost on the brewing industry (Brown 1916; Armstrong, 1921, 1937). The friendship between Hofmann and Böttinger had led to the employment, in the 1860s, of a number of Hofmann's students at Burton-on-Trent. At about the same time, both Böttinger and Hofmann returned to Germany. Böttinger died of cholera in 1872, but his son helped to turn Bayer into a major chemical company.

The group remaining at Burton included Horace Brown, Cornelius O'Sullivan and Peter Greiss. Greiss was mostly noted for his work on organic nitrogen chemistry and the synthesis of the azo dyes; O'Sullivan and Horace Brown published extensively on the chemistry and the microbiology of the brewing process. O'Sullivan defined the product of diastase as maltose, and not dextrose as had been confidently assumed. Brown's work on brewing was influenced by Pasteur, and he realized that the development of flavours in beers was due to secondary fermentations (Brown, 1916). Together they came close to anticipating the work of Michaelis and Menten in laying the foundations of enzyme kinetics (Boyde, 1980).

The science of microbiology in this period at Burton was as advanced as anywhere in Europe. It is interesting to reflect that in Britain, as well as in France, the stimulus for the work in biocatalysis was still the brewing industry. Brown deplored the title *Studies in Fermentation* given to the English translation of Pasteur's *Etudes sur la bière* (Pasteur, 1876):

How far this change of title was made with the consent of the author does not appear in the translator's preface, but, be that as it may, the alteration was a very unfortunate one, since it has helped to obscure the very close connection which exists between the subsequent developments of Pasteur's work in preventative medicine and these earlier studies on beer.
(Brown, 1916, p. 295)

There was a considerable interchange of ideas amongst the scientists working for the different brewers at Burton. An informal dining club developed into the Bacterium Club (Brown, 1916), whose meetings later (1887) were to be recorded as the *Transactions of the Laboratory Club*. In time this was to become the *Journal of the Institute of Brewing*. The contemporary literature suggests that the scientific climate at Burton was an exciting one. The relationship between the organic chemists and the brewing industry was very close.

1.8 Brown and the oxidation of mannitol

In 1873 Adrian Brown joined the group at Burton. He was Horace Brown's younger brother. He also had trained in London, at the Royal College of Science. Hofmann's Royal College of Chemistry had merged with the Royal College of Mines to form this new college, which was later to become Imperial College. After leaving the college, he spent a brief period at St. Bartholomew's Hospital, and then he returned to Burton to join one of the brewers, Salt & Co.

In 1886 Brown described the properties of two organisms, *Bacterium aceti* and *Bacterium xylinum*. He had isolated them from the surface of a

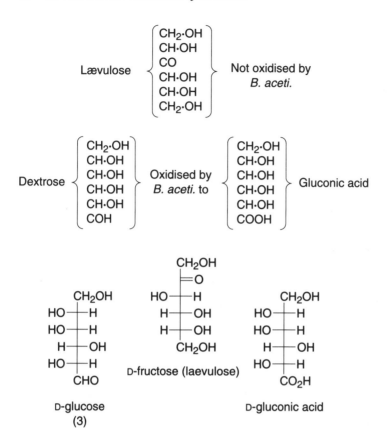

Scheme 1.5. The conversion and the "bracket" structures are reproduced directly from the original paper.

naturally contaminated beer whose alcohol had been oxidized to acetic acid. *B. aceti* oxidized ethanol and propanol, but, to his surprise, not methanol, nor isobutanol. Glucose was oxidized to gluconate (Scheme 1.5), a process which Boutroux had catalysed in 1880 with cultures of *Mycoderma aceti*, although the product had no detectable optical rotation (Brown, 1886a). In fact, genuine D-gluconate has a small rotation ($\alpha_D = +13°$). Finally, the organism oxidized mannitol to laevulose (fructose), at the time a quite novel reaction (Scheme 1.6). Brown thought that chemists and biologists alike would find the specificity of these reactions interesting, "as they help to show that the vital functions of certain organised ferments are most intimately connected with the molecular constitution of the bodies upon which they act" (Brown, 1886a, p. 187).

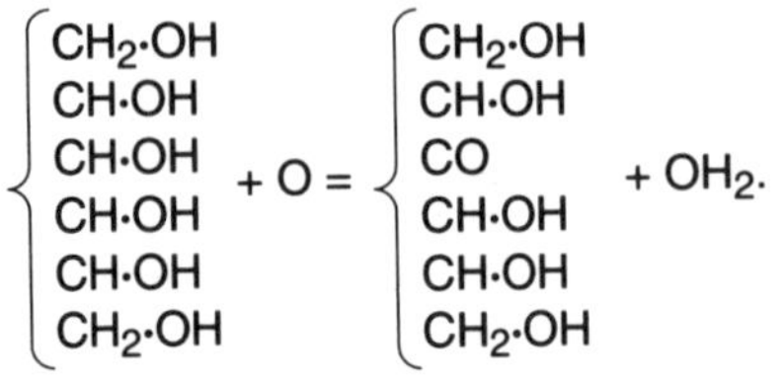

$$\left\{\begin{array}{l} CH_2\cdot OH \\ CH\cdot OH \\ CH\cdot OH \\ CH\cdot OH \\ CH\cdot OH \\ CH_2\cdot OH \end{array}\right. + O = \left\{\begin{array}{l} CH_2\cdot OH \\ CH\cdot OH \\ CO \\ CH\cdot OH \\ CH\cdot OH \\ CH_2\cdot OH \end{array}\right. + OH_2.$$

Mannitol   Lævulose

Scheme 1.6. Reproduced from the original publication.

The other organism, *B. xylinum*, also catalysed these reactions. He recognized it as similar to one of the organisms in the "vinegar plant", the other being *Saccharomyces cerevisiae*, and he found it to be responsible for the membrane of the "plant". Brown (1886b) had named the organism *B. xylinum* after showing that the membrane was made of cellulose, in contrast to the dextran of *Leuconostoc mesenteroides*. A century later the synthesis of cellulose in *B. xylinum*, now reclassified as *Acetobacter xylinum*, is still an active topic for research (Cannon and Anderson, 1991).

B. aceti also oxidized glycol (ethane-1,2-diol) to glycollate, and glycerol to carbonate, but, surprisingly, erythrol [erythritol (6)] was unaffected. Believing that sodium amalgam ought to reduce fructose, or indeed glucose, back to mannitol, he also commented that if the microbial oxidation were to follow the chemical reduction, it should be possible to convert glucose into fructose (Brown, 1887). Glucose was certainly reduced to mannitol, which he crystallized. The yield was low, about 7%, but the melting point, 164 °C, was good; pure mannitol melts at 168 °C. He then repeated the ealier experiment, oxidizing the mannitol with *B. aceti*, and recovered fructose.

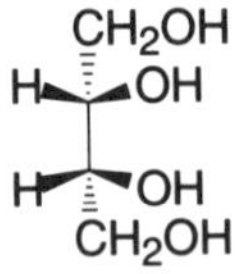

CH$_2$OH
H⎓OH
H⎓OH
CH$_2$OH

Erythritol (6)

This work again illustrates the problems of reading too much into the early literature. Even at the time, the experiment was thought unsatisfactory. Brown suggested that the six weeks needed for the mannitol to crystallize was the reason why there were doubts that this compound really was the product of the reduction of glucose. In fact, it would be 10 years before the chemistry of this simple experiment could be fully resolved.

A major advance came with detailed insight into the stereochemistry of sugars (Fischer, 1891). This crucial description clarified their chemical relationships, but would suggest that, in Brown's experiments, D-glucose (3) should be reduced to D-glucitol (7), not to D-mannitol (8). The former

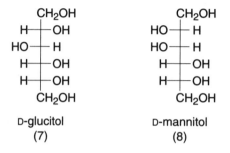

D-glucitol
(7)

D-mannitol
(8)

was certainly not the product which Brown had isolated, since its melting point, 110–112 °C, was too low. The explanation lies in the interconversion of aldoses and ketoses in alkaline solution (de Bruyn and van Ekerstein, 1895). Glucose is converted into a mixture of glucose, mannose and fructose; thus alkaline reduction with sodium amalgam produces a mixture of mannitol and glucitol (Scheme 1.7). This explains the low yield of crystals, due to the difficulty of obtaining mannitol from the mixture. Against the background of the contemporary state of sugar chemistry, the errors of interpretation are hardly surprising. In retrospect, it is curious that Brown did not record the optical rotations in these experiments, even though this was common practice at the time. Nor did he record any attempt to oxidize the mixture obtained from the alkaline reduction, rather than the mannitol obtained from it. This would have produced a mixture of D-fructose and L-sorbose (Scheme 1.7). Unfortunately, the notebooks describing these experiments, which would now be interesting to read, have long since vanished, probably during the wholesale reorganization of the brewing industry at Burton-on-Trent.

Nevertheless, the key observation (i.e. the oxidation of mannitol) remains unchallenged. During the 1890s Bertrand published a series of studies on an organism carried by a red vinegar fly which would oxidize sorbitol (D-glucitol) to sorbose. He defined the general reaction two years later as one in which the secondary alcohol oxidized was one of a pair of *cis*-hydroxyl groups adjacent to a primary hydroxyl group (Scheme 1.8). This general description, which was possible only against the background

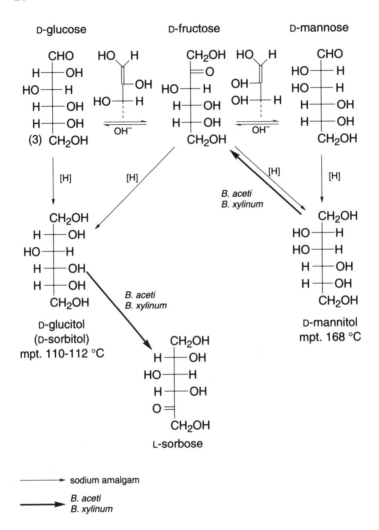

Scheme 1.7.

Scheme 1.8. General case for the oxidation performed by the "sorbose bacterium", *Acetobacter xylinum*.

of Fischer's chemistry, encompasses not only the oxidation of mannitol to fructose, which Brown had described, but, as Bertrand showed, also the oxidation of erythritol. Some years later Bertrand identified his organism, the sorbose bacterium, as Brown's *B. xylinum* (Bertrand, 1904), so fixing the link with the work in the breweries at Burton-on-Trent.

1.9 Fischer and the specificity of enzyme action

The same decade saw the development of a modern concept of enzyme action. Fischer published his studies of the hydrolyses of the α- and β-glycosides of D- and L-glucose in 1894 (Boyde, 1980; Teich and Needham, 1992). The active agents which he used to catalyse these reactions were the soluble ferments separated from *S. cerevisiae* and other micro-organisms, "the so-called enzymes", as he described them, using the term which Kühne had introduced in 1878 (Boyde, 1980). Some of these preparations catalysed a range of hydrolyses which do, in fact, reflect the action of several enzymes (Boyde, 1980); nevertheless, their specifications were sufficiently clear cut to allow him to propose

that enzyme and glycoside must match each other like lock and key in order to exert a chemical influence on each other. This concept appears more probable and of greater value for stereochemical research once the phenomenon itself is considered in purely chemical terms, rather than biological.
(Fischer, 1894, p. 2992; see Boyde, 1980)

The biochemistry of the period was in no position to acknowledge the complexity of these preparations of enzymes. Some years before, Brown (1886b) had even queried whether an individual cell of *B. xylinum* could both oxidize glucose to gluconate and at the same time assimilate it into cellulose. No more could Buchner, in 1897, imagine the nature of the cell-free system with which he converted sucrose into alcohol (Boyde, 1980).

If the remarkable variety of enzymatic catalysis could not be appreciated, then the application of its specificity to organic chemistry certainly was. In the same year as Buchner's publication, the Chemical Society in London arranged a Pasteur Memorial Lecture. It was delivered by Frankland, who was professor of chemistry at Mason College in Birmingham. In reviewing Pasteur's influence on organic chemistry and fermentation, he pointed out how Pasteur had anticipated Fischer's insight into enzymatic action. He also reviewed Pasteur's three methods for separating the enantiomers in a racemate. Their resolution with micro-organisms was, by then, a well-known process. He wrote that

it would be impossible for me here to attempt to place on record even the names of those numerous optically active compounds our acquintance with which is wholly dependent on the subjection of racemoids to the selective action of micro-organisms, indeed the method seems to be well nigh of universal application in the case of all racemoids bodies which are capable of being attacked by these low forms of vegetable life, and selective decompositions of this kind have been effected both by moulds, by bacteria and by the saccharomyces or yeasts.
(Frankland, 1897, p. 698)

These views, and the research on which they were based, were part of the mainstream of organic chemistry. The nature of enzyme action was also by this time an important topic for research in its own right. The developments which eventually led to the Michaelis-Menten theory were ones in which Adrian Brown, his elder brother Horace, and others from the group at Burton-on-Trent all played a major part (Boyde, 1980; Teich and Needham, 1992). Curiously, in their publications, the word "catalysis" itself is conspicuous by its absence.

1.10 Hill, Pottevin and the reversibility of enzyme action

Before the turn of the century, the word "catalysis" was rarely used, but nevertheless the essentially degradative aspects of Berzelius's original description of the process had been questioned. As far as enzymes were concerned, there was some doubt as to whether their action was reversible and as to whether the reactions in which they took part obeyed the law of mass action.

There is one report on the reversal of an enzymic (proteolytic) reaction dating back to 1886. Danilewski noticed that a heat-sensitive agent in stomach extracts would cause a concentrated peptic hydrolysate to coagulate. This reaction, whose product Sawjalow named "plastein" in 1901, was much studied between 1900 and 1925 (Wasteneys and Borsook, 1930). However, its chemistry remained poorly defined.

Hill (1898) showed the reversal of hydrolysis with a maltase extracted from a brewer's yeast. He dried the organism to a powder, which he then heated to $100\,^{\circ}\mathrm{C}$ while dry. The maltase was finally extracted from the yeast with 10 volumes of 0.1% (w/v) sodium hydroxide. The extract, which was neutral to litmus, not only hydrolysed maltose [α-D-Glc-$(1 \rightarrow 4)$-Glc (9)] to glucose, but also would synthesize a product from glucose which Hill believed, on the basis of a careful analytical study of the osazone, to be maltose. At a total sugar concentration of 40% (w/v) the reaction would approach its equilibrium position from either side (Table 1.4, Figure 1.2), and the synthesis was less effective at lower concentrations. Emmerling

Table 1.4. *Conditions for the reversal of maltose hydrolysis catalysed
by an extract of yeast cells*

Reaction conditions	Experiment			
	VIIIA	VIIIB	VIIIC	IX
Glucose (g)	9.81	3.98	19.65	2.95
Maltose H$_2$O (g)	—	—	—	0.98
Yeast extract (ml)	5	2 (boiled)	15	2
Total volume (ml)	25	10	50	10
Equivalent glucose concentration (g·ml^{-1})	3.9	4.0	3.9	3.9

Source: Data from Hill (1898).

identified the product of a similar reaction as isomaltose [α-D-Glc-(1 → 6)-
Glc (10)] rather than maltose.

maltose (9)

isomaltose (10)

Pottevin (1906) reviewed the evidence for the reversal of enzyme action
without any reference to the "plastein" reaction (Table 1.5). While he
accepted the strength of the evidence, he thought a test was needed in

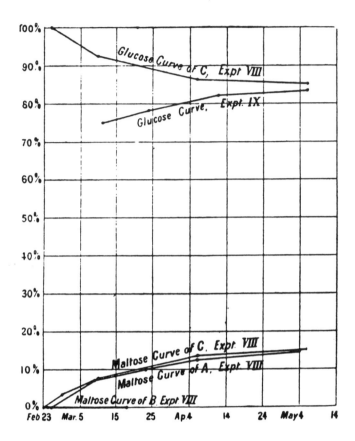

Figure 1.2. Reversal of maltose hydrolysis catalysed by an extract of dried yeast. The extract, prepared as described in the text, was mixed with solutions of glucose and maltose as described in Table 1.4. At various times (see dates) the amounts of maltose and glucose were estimated from the reducing power and the optical rotation of the solutions. For each experiment (see Table 1.4) the results are plotted as the percentage of the total carbohydrate present as glucose and maltose. (From Hill, 1898; reproduced with the kind permission of the Royal Society of Chemistry from the original paper, (Fig. 5, p. 652).)

which the product of the synthetic reaction would be the major component at equilibrium. He realized that the concentration of water was the main bar to effective synthesis in reactions which were essentially hydrolytic. He therefore decided to synthesize an ester in a reaction mixture from which water was, as far as possible, eliminated. He used a pancreatic extract and ran the reaction in organic solvent (Table 1.6, Figure 1.3). It was a classic demonstration of the synthetic use of an esterase in such conditions. The

Table 1.5. *Pottevin's list of reversible enzyme activities (1906)*

Substrate	Product	Enzyme	Reference[a]
Glucose	Isomaltose	Yeast maltase Takadiastase	Hill (1898) Emmerling (1901)
Butyrate & glycerol	Butyrin	Pancreatic extracts; serolipase	Hanriot (1901)
Glucose & mandelonitrile	Amygdalin	Yeast maltase Emulsin	Emmerling (1901) Fischer & Armstrong (1902)
Glucose & galactose	Isolactose	Lactase from Keffir grains	Fischer & Armstrong (1902)
Glucose & acetate	Glucosyl acetate	Pancreatic extracts	Acree & Hinkins (1902)
Glucose & fructose	Sucrose	Invertase	Arie & Wiser (1904)
Glucose & salicylate	Salicylin	Emulsin	Arie & Wiser (1904)
Butyrate & ethanol	Ethylbutyrate	Pancreatic extracts	Kastle & Loewenhardt (1904)

[a]For reference, see Pottevin (1906).

Table 1.6. *Conditions for the synthesis of methyl oleate catalysed by pancreatic diastase*

Reaction conditions	Exp. I	Exp. II	Exp. III
Oleic acid (g)	100	100	100
Methanol (g)	0	11.5	11.5
Water (g)	22	22	22
Tissue (g)	0	5 (heated)	5

Source: Data from Pottevin (1906).

yields of ester were high, but he did notice that a little water was an essential component of the reaction.

All of these reactions have their counterparts in modern biotechnology (see Chapters 3 and 6). The presence of these experiments at the turn of the century raises some interesting questions concerning the reasons for their subsequent consignment to the collective unconscious of biochemistry.

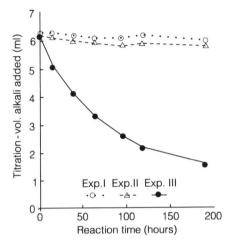

Figure 1.3. Synthesis of methyl oleate catalysed by pancreatic diastase. Porcine pancreas was minced under 95% ethanol and filtered. The residue was dried with ethanol and ether. Samples of the dried powder were tested as catalyst for the synthesis of methyl oleate as described in Table 1.6. At various times (hours) the amount of oleic acid remaining was titrated with alkali. The figure is drawn using the data in the original paper (Pottevin, 1906, p. 909).

Their relevance to our contemporary interests should not be anticipated from their distance of nearly a century. Once they had served their purpose in demonstrating that the reactions which enzymes catalysed were influenced by the effects of mass action, they had only limited relevance for the development of enzyme kinetics. In that context it is hardly surprising that they were forgotten.

1.11   Reichstein, Howarth and vitamin C

The preceding sections have traced the developments in microbiology, biochemistry and organic chemistry which allowed industry in the early twentieth century to make increasing use of biochemical technology in its manufacturing processes (Turner, 1994). Neidleman (1990) lists a number of important patents from that period; Takamine, Röhm, Wallerstein, Boidin and Effront all developed processes based on microbial amylase and proteases which replaced traditional methods of treating fabrics and leather and the brewing of beers. Microbial routes to butanol, acetone and glycerol were developed in response to shortages created by World War I; the work is associated with Weitzmann, Neuberg and Fernbach.

Currie pioneered the fermentation process for the production of citric acid. An interesting contemporary review of this work documents its scale and its promising future (Chapman, 1921).

By 1930, biocatalysis was also integrated into multistep chemical syntheses. The manufacture of D-ephedrine was based on Neuberg's demonstration that yeast would convert benzaldehyde to phenylacetyl-carbinol [(R)-1-phenyl-1-hydroxypropan-2-one (11)] (Neuberg and Ohle, 1922). Of particular interest, because of its relation with the work of Brown and Bertrand, is the synthesis of vitamin C [L-ascorbic acid (12)]. Two

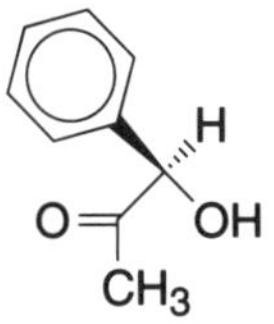

phenylacetylcarbinol
(R)-1-phenyl-1-hydroxypropane-2-one

(11)

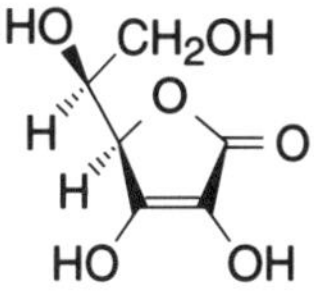

L-ascorbic acid
(vitamin C)

(12)

groups, one in Birmingham (Ault et al., 1933), the other in Zürich (Reichstein et al., 1933), separately published a chemical synthesis based on an input of L-xylosone (Scheme 1.9). The latter is itself difficult to synthesize and was no basis for a manufacturing process. Then, in the following year, Reichstein and Grössner (1934) published a productive synthesis (*eine ergiebige Synthese*) based on the oxidation of D-sorbitol (D-glucitol) to L-sorbose [Scheme 1.10; see oxidation of (13) to (14)]. The catalyst for this oxidation was Bertrand's sorbose bacterium, *B. xylinum*, and the reaction was the one which Brown had studied with D-mannitol

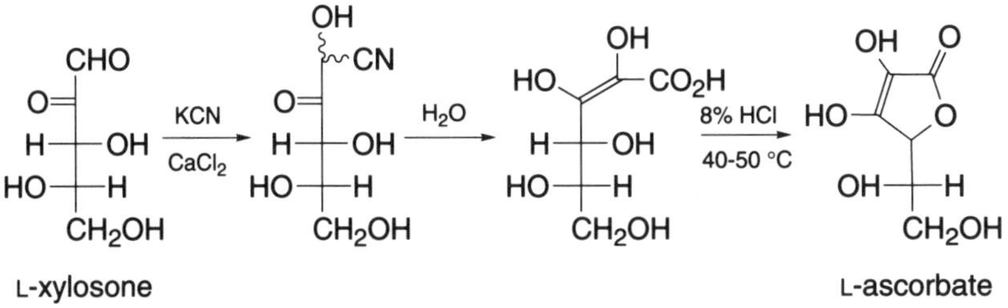

Scheme 1.9.

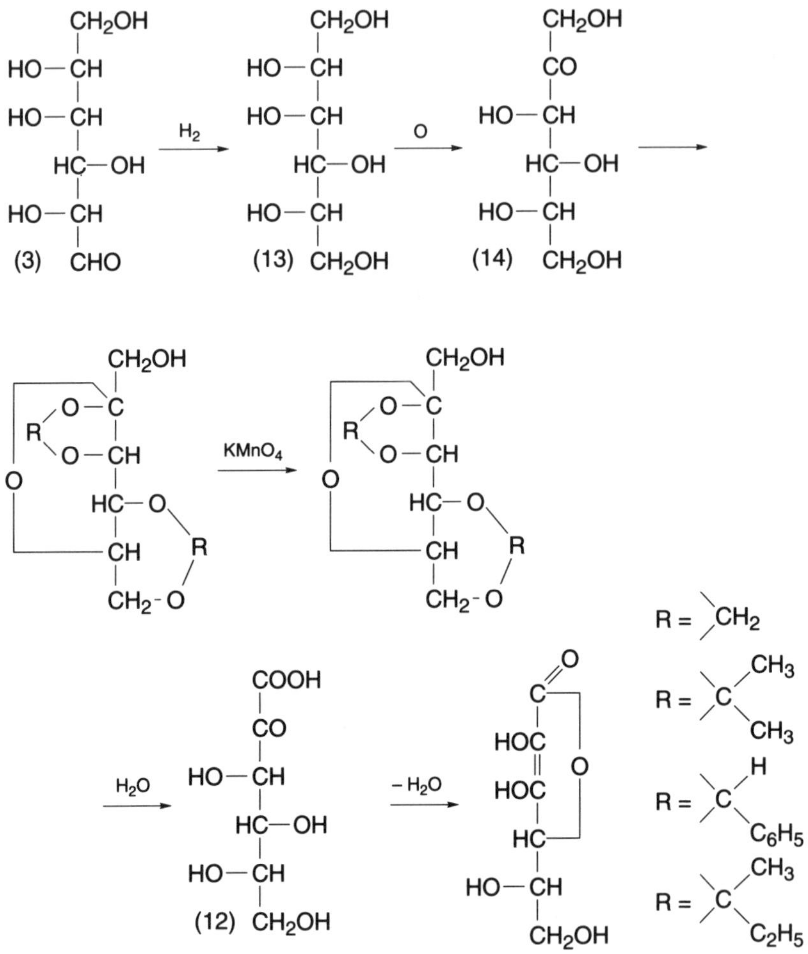

Scheme 1.10. Reproduced from the original publication.

as the substrate. It is not inconceivable that, amongst the products of alkaline reduction of glucose, he may also have oxidized D-glucitol (Scheme 1.7).

1.12 Conclusions and overview

This use of biocatalysis may come as a surprise to a generation of chemists and biologists who are struggling to accommodate a new technology called "biotransformations". It is pertinent now to rephrase the question posed at the beginning of this chapter and to ask why this science flourished 60 years ago (as exemplified by the production of vitamin C) and why we are now surprised by the resurgence of interest in this field.

Biotechnology exists at the interface among organic chemistry, biochemistry and chemical engineering. Through the nineteenth century this was the natural province for the brewing industry, and the fermentation process became more reliable and profitable as advances in all of these disciplines were applied to the conversion of sugar into alcohol. The wider application of fermentation technology outside of brewing evolved naturally. However, a split was about to develop which would hinder further progress.

Pasteur is now remembered as a founder of medical microbiology, although his contemporaries clearly had a broader view which included his studies on brewing and organic chemistry. There is little doubt that, compared to the nineteenth-century chemist, the twentieth-century biochemist has moved towards an interest in medicine and physiology. There is enough evidence of that shift in the papers which Teich and Needham (1992) have collected together. Organic chemists also set out on a different path. The two disciplines began to pursue separate goals; broadly defined, biochemistry investigated living processes, while organic chemistry concentrated on analysis and synthesis. This was essential for the two sciences to make the rapid progress which has been achieved in the twentieth century.

A study of Adrian Brown's career after he left Burton-on-Trent in 1899 is also instructive. He took the chair of brewing and malting at Mason College in Birmingham, a position which the brewers had newly endowed. Later the Royal Institue of Chemistry appointed him as its first examiner in biological chemistry. Finally, at Birmingham he became professor of the biology and chemistry of fermentation, and director of the School of Brewing.

Some of his research during this period (Armstrong, 1921) concerned the rate of enzyme action, and his work was cited when Michaelis and Menten published their definitive theory in 1913. He was, however, some way from anticipating their treatment of the kinetics, except in clearly recognising the intermediate formation of an enzyme-substrate complex, and the effect this would have on the rate of reaction. This process would explain the apparent deviations in the rate from a law of mass action based on the reagent concentrations alone (Brown, 1902). He was elected to the Royal Society in 1911, and he died in 1919, a greatly respected member of the university (Plate 1.1).

Plate 1.1. Portrait of Adrian Brown. (From Armstrong, 1921; reproduced with the kind permission of the Institute of Brewing.) Memorial plaque now in the School of Biochemistry at the University of Birmingham. (Reproduced with the kind permission of the head of the School of Biochemistry, University of Birmingham.)

The memorial plaque is not the only tribute to the early history of bio-catalysis at Birmingham. The Frankland Building commemorates Percy Frankland, who was professor of chemistry at Mason College, and whose views of biocatalysis have already been noted (Frankland, 1897). The Howarth Building commemorates the leader of the group at Birmingham responsible for their synthesis of L-ascorbic acid (Scheme 1.9).

Is it significant that the two chairs which Adrian Brown held, professor of brewing and malting and professor of the biology and chemistry of fermentation, suggest the beginning of the specialization which has been a prerequisite for the advances made in twentieth-century science? The broad approach to organic chemistry and microbiology, which would have been as familiar to many other groups as it must have been to this one at Birmingham, could not survive for long. With it went the integrated structure which had laid the foundation for the manufacture of L-ascorbate and for some other important industrial processes. It became inevitable that the organic chemist would eventually find biological catalysis un-familiar ground and that organic synthesis would become equally unfamiliar to the biochemist.

The recent advances, in the latter part of the twentieth century, have come as the two disciplines, gaining from their hugely increased knowledge, identify problems to solve together. However, the knowledge base is now so large that this convergence and integration of the disciplines is unlikely to be achieved by individuals working alone. It is more likely to be achieved by groups of individuals effecting the same synergy among organic chemistry, biochemistry and engineering which was so effective in the last century. Such collaborative work can take advantage of the knowledge which the past 60 years have given us. These are the subjects of subsequent chapters.

References

Armstrong, H. E. (1921). Adrian Brown Memorial Lecture. The Particulate Nature of Enzymic and Zymic Change. *J. Inst. Brewing, 18*, 197–260.

Armstrong, H. E. (1937). Horace Brown Memorial Lecture. *J. Inst. Brewing, 43*, 375–86.

Ault, R. G., Baird, D. K., Carrington, H. C., Haworth, W. N., Herbert, R., Hirst, E. L., Percival, E. G. V., Smith, F., & Stacey, M. (1933). Synthesis of *d*- and of *l*-Ascorbic Acid and of Analogous Substances. *J. Chem. Soc.*, 1419–23.

Beral, M. P.-J. (1815). Notes sur la fermentation. *J. de Pharmacie et des Sciences Accessoires (Paris), 1*, 358–61.

Bertrand, G. (1904). Etude biochemique de la bactérie du sorbose. *Annales de Chemie et de Physique (Paris)*, 8me. série *3*, 181–288.

Berzelius, J. J. (1838). *Traité de Chimie*, translated from the 4th German edition (1838) by Valerius, B. Bruxelles. Société Typographique Belge, Adolphe Wahlen & Cie.

Bouillon-Lagrange (1811). Sur le passage de l'amidon à l'état de muqueux, et sur quelques teintures noires. *Bulletin de Pharmacie (Paris) 3*, 395–8.

Boyde, T. R. C. (1980). *Foundation Stones of Biochemistry.* Voile et Aviron, Hong Kong.

Brown, A. J. (1886a). Chemical Action of Pure Cultures of *Bacterium aceti. J. Chem. Soc., 49*, 172–87.

Brown, A. J. (1886b). On an Acetic Acid Ferment which Forms Cellulose. *J. Chem. Soc., 49*, 432–9.

Brown, A. J. (1887). Futher Notes on the Chemical Action of *Bacterium aceti. J. Chem. Soc., 51*, 638–42.

Brown, A. J. (1902). Enzyme Action. *J. Chem. Soc., 81*, 373–400.

Brown, H. T. (1916) Reminiscences of Fifty Years' Experience of the Application of Scientific Method in Brewing Practice. *J. Inst. Brewing, 13*, 265–354.

Cannon, R. E., & Anderson, S. M. (1991). Biogenesis of Bacterial Cellulose. *Crit. Rev. Microbiol., 17*, 435–47.

Chapman, A. C. (1921). Microorganisms and Their Uses. *J. Roy. Soc. Arts, 69*, 581–9, 597–605, 609–19.

de Bruyn, C. A. L., & van Ekerstein, W. A. (1895). Action of Alkalis on Sugars. Reciprocal Transformation of Glucose, Fructose and Mannose. *Ber., 28*, 3078–3082; see also *J. Chem. Soc.* (1896), *70*, 116–17.

De Saussure, T. (1814). Sur la conversion de l'amidon. *Bulletin de Pharmacie (Paris), 6*, 499–504 (abstract).

Dessaignes (1849). Note sur la conversation du malate de chaux en acide succinique. *Annales de Chimie et de Physique (Paris)*, 3me. série *25*, 253–5.

Döbreiner (1815). Expériences sur le ferment. *J. de Pharmacie et des Sciences Accessoires (Paris)*, *1*, 342–5.

Fischer, E. (1891). Configuration of Grape Sugar and Its Isomerides. *Ber.*, *24*, 1836–45; see also (i) *J. Chem. Soc.* (1891) *60*, 1173–8, 1444–7, and (ii) Streitweiser, A., & Heathcock, C. H. (1985). *Introduction to Organic Chemistry*, 3rd ed. Macmillan, New York.

Fischer, E. (1894). Effect of Configuration on Enzyme Activity. *Ber.*, *27*, 2985–93.

Frankland, E. (1885). On Chemical Changes in Their Relation to Microorganisms. *J. Chem. Soc.*, *47*, 159–83.

Frankland, P. (1897). Pasteur Memorial Lecture. *J. Chem. Soc.*, *71*, 683–743.

Guérin-Varry, R. T. (1835). Memoire concernant l'action de la diastase sur l'amidon de pommes de terre. *Annales de Chemie et de Physique (Paris)*, 2me. série *60*, 32–78.

Hill, A. C. (1898). Reversible Zymohydrolysis. *J. Chem Soc.*, *73*, 634–58.

Hofmann, A. W. (1861). On the Importance of the Study of Chemistry. In *Lectures Addressed to Teachers, Lecture 7*. Science and Art Department of the Committee of Council on Education, London.

Hofmann, A. W., & Graham, T. (1852). Report upon the Alleged Adulteration of Pale Ales by Strychnine. London.

Jorpes, J. E. (1966). *Jac. Berzelius: His Life and Work*. Almqvist & Wiksell, Stockholm.

Kirchhoff (1816). Formation du sucre. *J. de Pharmacie et des Sciences Accessoires (Paris)*, *2*, 250–8.

Klein, E. (1886). Bacteriological Research from a Biologist's Point of View. *J. Chem. Soc.*, *49*, 197–205.

Leaback, D. (1992). What Hofmann Left Behind. *Chem. Ind.*, 378.

Leibig, J., & Kopp, H. (1849). *Annual Report of Progress in Chemistry*, vol. 3, ed. A. W. Hofmann & H. Bence-Jones, pp. 207–8. Taylor, Walton & Maberley, London (1852).

Neidleman, S. L. (1990). The Archeology of Enzymology. In *Biocatalysis*, ed. D. A. Abramowicz, pp. 1–24. Van Nostrand Reinhold, New York.

Nettleton, J. A. (1893/4). The Flavour of Whisky – as Influenced by the Materials Used and by the Process of Manufacture. *Trans. Inst. Brewing*, *7*, 176–95.

Neuberg, C., & Ohle, H. (1922). Biosynthetic Carbon Chain Union in Fermentation Processes. *Biochem. Z.*, *128*, 610–18.

Packowski, G. I. (1978). In *Kirk-Othmer Encyclopedia of Chemical Technology*, 3rd ed., vol. 3, ed. M. Grayson & D. Eckroth, p. 851. Wiley, New York.

Pasteur, L. (1858a). Memoire sur la fermentation de l'acide tartrique. *C. R. Acad. Sci. (Paris)*, *46*, 615–18.

Pasteur, L. (1858b). Nouvelles recherches sur la fermentation alcoolique. *C. R. Acad. Sci. (Paris)*, *47*, 224.

Pasteur, L. (1860). Note relative au *Penicillium glaucum* et a la dissymétrie moléculaire des produits organiques naturels. *C. R. Acad. Sci. (Paris)*, *51*, 298–9.

Pasteur, L. (1876). *Etudes sur la bière, ses maladies, causes qui les provoquent, procédé pour la rendre inaltérable, avec une théorie nouvelle de la fermentation*. Gauthier-Villars, Paris. Translated into English (1879) as *Studies in Fermentation, the Diseases of Beer, Their Causes, and the Means of Preventing Them*, ed. F. Faulkner, & D. C. Robb, Macmillan, London.

Payen, A. & Persoz, J. F. (1833). Mémoire sur la diastase, les principaux produits de ses reactions, et leurs applications aux arts industriels. *Annales de Chemie et de Physique (Paris)*, 2me. série *53*, 73–92.

Phipson, T. L. (1862). On the Transformation of Citric, Butyric and Valeric Acids with Reference to the Artificial Production of Succinic Acid. *J. Chem. Soc.*, *15*, 141–2.

Playfair, L. (1848). On Transformation Produced by Catalytic Bodies. *Mem. Proc. Chem. Soc. London*, *3*, 348–70.

Plimpton, R. T. (1881). On the Amylamines Corresponding to the Active and Inactive Alcohols of Fermentation. *J. Chem. Soc.*, *39*, 331–6.

Pottevin, H. (1906). Actions diastasiques reversibles. Formation et dédoublement des éthers-sels sous l'influence des diastases du pancréas. *Annales de l'Institute Pasteur*, *20*, 901–23.

Reichstein, T., Grössner, A., & Oppenauer, R. (1933). Synthese der *d*- und *l*-ascorbinsäure (C-vitamin). *Helv. Chim. Acta*, *16*, 1019–33.

Reichstein, T., & Grössner, A. (1934). Eine ergiebige Synthese der l-ascorbinsäure (C-vitamin). *Helv. Chim. Acta*, *17*, 311–28.

Root-Bernstein, R. S. (1989). *Discovering*. Harvard University Press.

Sykes, W. J. (1895). The Indebtedness of Brewers to M. Pasteur. *J. Federated Inst. Brewing*, *1*, 498–525.

Teich, M., & Needham, D. M. (1992). *A Documentary History of Biochemistry 1770–1940*. Leicester University Press.

Travis, A. S. (1992). The Man Who Put Science into Industry. *Chem. Ind.*, 302.

Turner, M. K. (1994). Biological Catalysis and Biotechnology. In *The Chemical Industry*, 2nd edition, C. A. Heaton, pp. 306–71. Blackie, Glasgow.

Vallery-Radot, R. (1901). *The life of Pasteur* (trans. R. L. Devonshire). Constable, London.

Vogel (1812). D'experiences sur la fabrication du sucre d'amidon. *Bulletin de Pharmacie (Paris)*, *4*, 255–6 (abstract).

Vogel (1817). Sur la formation de l'acide lactique pendant la fermentation. *J. de Pharmacie et des Sciences Accessoires (Paris)*, *3*, 491–3.

Wasteneys, H., & Borsook, H. (1930). The Enzymatic Synthesis of Proteins. *Physiol. Rev.*, *10*, 110–45.

2

The interrelationships between enzymes and cells, with particular reference to whole-cell biotransformations using bacteria and fungi

2.1 Introduction

There are basically two strategies for carrying out biotransformations:
(1) to use pure or partially purified enzymes isolated by the investigator or purchased from a commercial supplier, or (2) to use whole cells. Enzymes are categorized by the Enzyme Commission (EC) according to their functions (Table 2.1), and each individual enzyme is given a unique code made up of four numbers, such as 2.1.2.4. These reference numbers are derived as follows:

The first number indicates the class (1 through 6; see Table 2.1).

The second number in the series indicates the subclass. For oxido-reductases, the subclass number indicates the type of group in the donor which undergoes oxidation (1 denoting a secondary alcohol group, 2 denoting an aldehyde or ketone unit, etc.); for transferases, it gives an indication of the functional group which is transferred (1 indicates the transfer of a one-carbon unit); for the hydrolases, it earmarks the functional group hydrolyzed (1 is used when an ester group is hydrolyzed); for the lyases, it indicates the group HX (3 indicates the addition of ammonia); for the isomerases, it shows the type of isomerization (2 indicates alkene *cis–trans* isomerization); for ligases, it indicates the type of bond formed (4 indicates carbon–carbon bond formation).

The third number in the series serves to allocate the enzyme to a sub-subclass. For oxidoreductase enzymes, this third number shows the type of acceptor involved [e.g. 1 denotes a co-enzyme, such as nicotinamide-adenine dinucleotide phosphate (NADP); 2 denotes a cytochrome; 3 denotes molecular oxygen]. For the transferases, the third number allows a subdivision of the type of group transferred (thus the C$_1$ unit can be defined as methyl or carboxyl, etc.). For the hydrolases, the lyases, the isomerases, and the ligases, the third number shows more precisely the

Table 2.1. *Enzyme classification (EC system)*

Category	Examples of enzymes useful in synthesis
Oxidoreductases Interconvert ketones with alcohols, double bonds with single bonds, etc.	yeast/horse liver alcohol dehydrogenase; oxygenases
Transferases Transfer acyl, phosphoryl, sugar, amino groups, etc.	transaminases; kinases
Hydrolases Hydrolysis of esters, peptides, glycerides, anhydrides, etc.	lipases; esterases; acylases; proteases; phosphatases; glycosidases
Lyases Addition to double bonds $C=C$, $C=N$, $C=O$, etc.	$(C=O)$ aldolases; mandelonitrile lyase $(C=C)$ aspartase; fumarase
Isomerases Various Isomerizations, $C=C$ bond migration, *cis–trans* racemizations, etc.	fructose-glucose isomerase
Ligases Formation of $C—O$, $C—S$, $C—N$, and phosphoryl bonds	very important in molecular biology

type of bond hydrolyzed, the nature of the added group, further detail of the isomerization, and the type of substance formed, respectively.

The fourth number in the series completes the serial number for the enzyme, allowing each enzyme to have a unique four-digit number for reference purposes.

A key to the numbering and classification of the enzymes can be found in any text-book on biochemistry. The numbers of the different types of enzymes that have been identified and the numbers of enzymes that are commercially available are listed in Table 2.2 (these data are from 1990 texts).

An experimentalist can therefore have the choice of performing a biotransformation using an isolated enzyme or a whole-cell system. The advantages and disadvantages are summarized in Table 2.3, further details of which are discussed in Section 2.2. It is vitally important not to consider whole-cell biotransformation as a "black art", and so considerable effort has been made in the next sections to interrelate the activities of whole

Table 2.2. *Numbers of enzymes identified and numbers*
of enzymes commercially available

Enzyme type	No. of enzymes identified	No. of enzymes commercially available
Oxidoreductase	650	90
Transferase	720	90
Hydrolase	636	125
Lyase	255	35
Isomerase	120	6
Ligase	80	5

Table 2.3. *Some advantages and disadvantages of using whole-cell systems*
as opposed to enzymes

Biotransformation system	Advantages	Disadvantages
Whole cells	Inexpensive Enzyme co-factors present	Large glassware required Messy work-up Side reactions can interfere or dominate substrate and/or product, and/or co-solvent may disrupt (membrane-bound) enzymes
Isolated enzymes	Simple apparatus Simple work-up Specific for selected reaction Co-solvents better tolerated	Expensive Addition of enzyme co-factors required or enzyme co-factor recycling necessary

cells and enzymes, as well as to explain in detail the methodology behind whole-cell biotransformations using bacteria and fungi.

2.2 Interrelationships between enzymes and cells: choosing the best biotransformation system

The enzymes useful for undertaking biotransformations are themselves synthesized by the cells of living organisms. The cell is the fundamental unit of life. All living organisms consist of one or more cells – small membrane-bound units containing highly organized arrays of numerous molecules. Unicellular living organisms, typified by many species of bacteria, consist of a single cell type, whereas multicellular living organisms consist of an organized assemblage of a number of similar cells or, more usually, dissimilar cells. Each cell type produces its own characteristic complement of very many different enzymes, present in greater or lesser concentrations depending on the nature of the enzyme, the type of cell, and its stage of development. All enzymes capable of being produced from the genetic information encoded in the genome of a cell are synthesized within the confines of the peripheral plasma membrane. Whereas the majority of the enzymes synthesized by cells are retained to function intracellularly, there are some so-called extracellular enzymes that are subsequently secreted outside the boundary membrane of the cellular site of synthesis.

The normal role of the enzymes produced by a cell is to promote metabolism. Metabolism is the interactive sequence of biochemical reactions that collectively result in the growth and development of the cell. However, few of the many hundreds of different enzymes normally present in a typical cell are specific enough to catalyze reactions with only the evolved natural substrates that serve growth and development roles in the cell. As a generalization, the enzymes of catabolism (energy-yielding metabolism) have a broader substrate range than the enzymes of anabolism (biosynthetic metabolism). It is this catholic nature of the substrate specificity of many cellular enzymes that widens the scope of biotransformation that it is possible to undertake. Enzymes or, indeed, the cells of living organisms, most usually the cells of particular groups of microorganisms (bacteria, yeasts, and filamentous fungi), can be utilized.

This brief overview begs two questions with regard to biotransformations:

1. Why use whole cells themselves rather than the relevant enzyme complement?
2. Why use microbial cells rather than the cells of animals and plants?

The answer to the first question (Why use whole cells rather than isolated enzymes?) stems from problems associated with using some enzymes, especially intracellular enzymes (i.e. enzymes that normally function within the cells in which they are synthesized). Principal amongst these problems are the following:

A. Many potentially useful enzymes are relatively unstable outside the cellular environment in which they normally operate. Consequently such enzymes have little values as agents for biotransformation if extracted from source cells. The molecular basis of this instability and concomitant loss of catalytic activity is the loss of the conformational protein structure of the active forms of such enzymes. There are several reasons for this loss of the essential three-dimensional shape of intracellular proteins outside the cellular environment in which they normally exist. Many intracellular enzymes are maintained in restricted but highly active conformations by association to a greater or lesser extent with specific membrane-defined environments within cells, and such multimacromolecular coacervates are disrupted or destroyed by procedures designed to ensure the extraction of intracellular enzymes. This is particularly pertinent for many of the enzymes present in eukaryotic cells, which characteristically contain many intracellular enzymes in a number of different types of membrane-bound subcellular organelles. By way of contrast, some isolated enzymes lose activity as a result of the intermolecular aggregation of individual enzyme molecules that are normally maintained by membranes as discrete entities in the cells of living organisms. Methodologies developed to counter such problems include those based on providing some form of mechanical support for isolated enzymes (immobilization techniques) and those based on restricting conformational mobility by minimizing the peripheral hydration layer of macromolecules (enzymes and PEG-modified enzymes in apolar organic solvents).

B. Many potentially useful intracellular enzymes are subject to hydrolytic attack by other enzymes – particularly proteases – released during procedures to fractionate living cells. Proteases are essential for various aspects of growth and development and consequently are characteristic activities always present in cells.

C. Whereas some intracellular enzymes (e.g. hydrolases) are not dependent on one or more different types of co-factors to promote catalytic activity, many potentially useful intracellular enzymes (e.g. monooxygenases, oxidoreductases, ligases) are co-factor-dependent, for example nicotinamide nucleotide- and/or adenosine nucleotide-dependent:

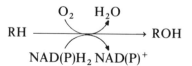

Such co-factors are recycled by complementary enzyme-catalysed reactions in living cells:

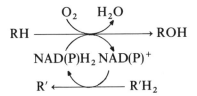

However, attempting to use a co-factor-dependent enzyme outside whole cells means either using expensive co-factors as co-substrates in stoichiometric amounts or developing some form of compatible in vitro co-factor recycling system. Although this has been achieved for some isolated enzymes (notably some oxidoreductases), this is not always possible and/or cost-worthy.

D. Preparation of cell-free intracellular enzymes can be expensive in terms of both time and resources, because the necessary steps in extraction and purification of enzymes from cells often are laborious and result in low overall yields of active enzyme preparations.

Note that whereas the foregoing criticisms do not apply to most extra-cellular enzymes (i.e. proteins subsequently secreted outside the cellular site of synthesis), these particular biocatalysts are almost exclusively hydro-lases of one sort or another (e.g. proteases) which have specific, and hence limited, potential uses in biotransformation.

The answer to the second question (Why use microbial cells?) is that the cells of micro-organisms are more desirable than those of animals and plants in several respects:

A. Because of the small sizes of many microbial cells, especially bacterial cells, compared with those of plants and animals, micro-organisms are able to grow relatively quickly. This is because small cells have a high surface-area:volume ratio, which in turn influences the potential to take up nutrients from the environment. A necessary corollary to promote rapid growth is that microbial cells have high rates of cellular metabolism, leading to fast rates of biotransformation of substrates undertaken by whole cells of micro-organisms.

B. As a generalization, the range of substrates that can be metabolized by microbial cells, especially the cells of some bacteria, is more extensive than the ranges for plant and animal cells. For instance, all species of fluorescent pseudomonads are nutritionally highly versatile, each being able to use 60 to 80 different organic compounds as sole source of carbon and energy.

C. As a generalization, microbial cells are far easier to grow in culture than are plant and animal cells. This is in part a reflection of the combination of small size and effective cell wall structure which gives many different types of microbial cells relatively high mechanical strength. This in turn makes them more resistant than animal and plant cells to the various rigours of different culture techniques. There are, however, some types of micro-organisms (e.g. obligate anaerobes and dangerous pathogens) that requires special growth facilities, thus making them unsuitable for use as agents for biotransformation except in laboratories equipped with the requisite specialist facilities.

While still considering microbial whole-cell biotransformation in conceptual terms, it is advisable to ponder the implications of one other important point, namely, that such processes exploit the inherent metabolic potential of the cells of biocatalysts. The cells of all living organisms, including micro-organisms, represent aqueous-based, enclosed microenvironments that have evolved to grow successfully by exploiting natural resources available from the biosphere, which is itself a predominantly water-based macro-environment. In most cases good growth requires that micro-organisms be cultured in aqueous-based liquid media. Consequently, unless appropriate measures are taken, there may be difficulties in undertaking biotransformation on highly lipophilic substrates, whereas hydrophilic products can also be troublesome to extract from spent reaction mixtures. In addition, xenobiotic compounds may prove to be toxic to cells above a relatively low threshold concentration. An allied problem is that those xenobiotic compounds that are capable of being biotransformed often result in dilute aqueous solutions of product(s). It is possible with some micro-organisms to overcome a number of these potential problems by undertaking biotransformation in aqueous:solvent biphasic environments, but this approach should be considered only by those with considerable experience of bioconversions.

2.3 Practical experimental methods for whole-cell biotransformations using bacteria and fungi

Undertaking small-scale microbial whole-cell biotransformations to the extent of being able to fully characterize the nature of the resultant products (including relevant aspects of stereochemistry where applicable) is a multidisciplinary exercise that requires the interaction of expertise and facilities most usually associated with the sciences of microbiology and chemistry. A microbiologist lacking the relevant chemical training and facilities is seriously disadvantaged, as is an ill-equipped chemist entering the world of microbiology. The best policy in these circumstances is to establish a working partnership between mutually interested chemists and microbiologists. Such a collaborative approach is an effective means to rapidly undertake initial probing reactions. (Is a particular molecule capable of being biotransformed by a certain micro-organism? If so, what products are formed?) The scope and extent of the collaborative effort can most significantly be enhanced if the participants can abandon the confines of their individual disciplines and spend a few days working in the complementary laboratory when relevant work is being actively pursued. This is the most effective method of gaining a practical grasp of the merits and demerits of the various experimental methodologies and the scale of investment in infrastructure required to operate at a particular level of technology.

From a pragmatic viewpoint, only those biotransformations that result in the majority of the desired product accumulating extracellularly should be considered practical propositions by experimenters other than those experienced in this area of biotechnology. Interestingly, the majority of biotransformations meet this criterion.

Set out in the following subsections is an outline of the contribution that microbiology can make to undertaking small-scale whole-cell biotransformations. For those unfamiliar with the basic principles of maintaining and growing micro-organisms, an outline of the fundamental techniques is given in Appendix A. Because this book is intended as an introductory text for scientists from different disciplines, the methodologies are not dealt with comprehensively, but rather in sufficient detail to draw attention to salient features that may in some cases require further investigation using a good conventional text-book of microbiology. Conversely, several aspects are detailed that either are not included or are taken as read in such books. A number of the illustrative examples are drawn from the use of microbial whole-cell biotransformations to accomplish Baeyer-Villiger mono-oxygenase-dependent ring expansion of alicyclic ketones to equivalent lactones (Scheme 2.1).

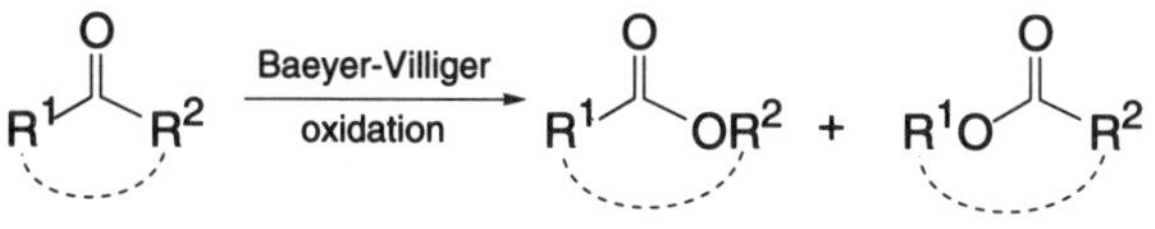

Scheme 2.1

In order to accomplish any small-scale microbial whole-cell biotransformation, many, if not all, of the following key questions need to be resolved:

1. What is a suitable micro-organism to act as the biocatalyst for a particular desired biotransformation?
2. What is the most suitable form of the chosen biocatalyst to use?
3. What needs to be done to maximize the relevant enzyme titre(s) of the chosen biocatalyst?
4. What, if anything, needs to be done to facilitate biocatalyst–substrate interaction?
5. What, if anything, needs to be done to optimize product yield? This will include means of preventing metabolic fates other than the required biotransformation for the substrate and/or product(s).

Each of these facets will be dealt with, in turn, in the following subsections (2.3.1–2.3.5).

2.3.1 What is a suitable micro-organism to act as the biocatalyst for a particular desired biotransformation?

The most useful form of microbial whole-cell biocatalyst is a pure culture of a micro-organism (i.e. a single species uncontaminated with other species) that contains high levels of an enzyme (or enzymes) that will carry out the desired bioconversion. The alternative possibility of using a mixed culture of two or more different types of micro-organisms, one of which will carry out the desired bioconversion, is not recommended because mixed cultures are often unstable, due to the changing relative population dynamics of the partners. This can result in unpredictable variation in the outcome of a biotransformation.

One rational way to seek such a pure culture of a biocatalyst is to screen the published literature for precedent examples of micro-organisms reported to undertake the desired type of biotransformation with structurally related molecules. In the Bibliography, in addition to original papers published in research journals, a number of books and review articles that detail such micro-organisms have been compiled for various groups of compounds. Considering, as a specific example, the search for micro-organisms potentially able to undertake Baeyer-Villiger mono-oxygenase-dependent ring expansion of alicyclic ketones to equivalent lactones, such a literature screen reveals many examples of micro-organisms reported to catalyse one or more equivalent reactions. Some of these micro-organisms, which include species of both fungi and bacteria, are listed in Table 2.4. Such a culture often can be obtained from the investigator who initially recorded its characteristic feature of interest. Alternatively, the

Table 2.4. *Microbial Baeyer-Villiger mono-oxygenases in lactone formation*

Substrate	Micro-organism	Reference
Borneol	*Pseudomonas pseudomallei*	Hayashi et al. (1970)
Camphene	*Pseudomonas camphene*	Khanchandani and Bhattacharaya (1974)
Camphor	*Mycobacterium rhodochrous* NCIMB 9784	Kay et al. (1962)
	Pseudomonas putida NCIMB 10007	Conrad et al. (1961)
1,8-cineole	*Pseudomonas flava* UQM 1724	Williams et al. (1989)
Cycloheptanone	*Nocardia* sp. KUC-7N	Hasegawa et al. (1982)
Cyclohexanone	*Pseudomonas* sp.	Tanaka et al. (1977)
Cyclohexane	*Nocardia* sp.	Stirling et al. (1977)
	Pseudomonas sp.	Anderson et al. (1980)
	Xanthomonas sp.	Trower et al. (1985)
Dihydrocarvone	*Acinetobacter* sp. NCIMB 9871	Abril et al. (1989)
Fenchone	*Mycobacterium rhodochrous* NCIMB 9784	Chapman et al. (1963)
2-hexylcyclopentanone	*Pseudomonas oleovoroans* NCIMB 6576	Shaw (1966)
Menthol	*Rhodococcus* sp. M-1	Shulka et al. (1987)
Menthone	*Pseudomonas putida* YK-2	Nakajima et al. (1978)
7-methyl-1,5-dioxo-4-indapropionic acid	*Nocardia restritus* ATCC 14887	Lee and Sih (1967)
Norbornanone	*Acinetobacter* sp. NCIMB 9871	Abril et al. (1989)
α-terpineol	*Mycobacterium rhodochrous* NCIMB 9784	Baum and Gunsalus (1962)
2,2,5,5-tetramethyl-1,4-hydroxycyclohexane	*Curvularia lunata* NRRL 2380	Ouazzani-Chadhi et al. (1987)
2-undecylcyclopentanone	*Acinetobacter* sp. NCIMB 9871	Alphand et al. (1990)
	Acinetobacter sp. TD63	

fact that some of these micro-organisms carry designations such as *Mycobacterium rhodochrous* NCIMB 9784, *Curvularia lunata* NRRL 2380, and *Aspergillus parasiticus* ATCC 15517 is important. These numbers indicate that these particular micro-organisms are deposited, respectively, with the National Collections of Industrial and Marine Bacteria (NCIMB), Aberdeen, England, the Northern Regional Research Laboratory (NRRL), Peoria, Illinois, USA, and the American Type Culture Collection (ATCC),

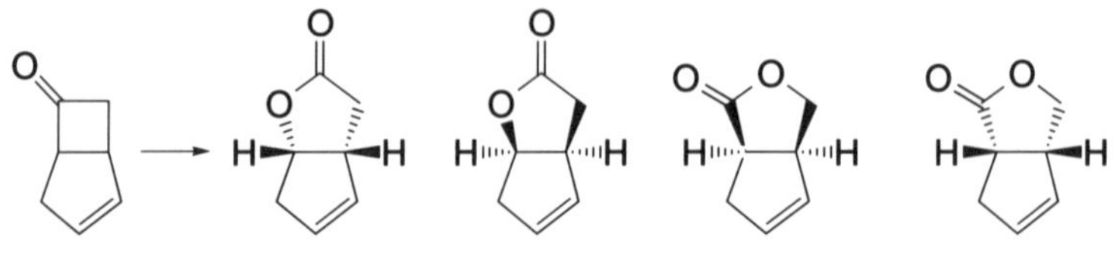

| racemic bicyclo (3.2.0.) hept- 2-en-6-one (±) 1 | (−)-(1S,5R)- 2-oxabicyclo (3.3.0.) oct-6 -en-3-one (−) 2 | (+)-(1R,5S)- 2-oxabicyclo (3.3.0.) oct-6 -en-3-one (+) 2 | (−)-(1R,5S)- 3-oxabicyclo (3.3.0.) oct-6 -en-2-one (−) 3 | (+)-(1S,5R)- 3-oxabicyclo (3.3.0.) oct-6 -en-2-one (+) 3 |

Figure 2.1. Biotransformation of bicyclo(3.2.0)hept-2-en-6-one by fungi.

Rockville, Maryland, USA, which are all open-access culture collections. As such, these micro-organisms are available on request (and payment of a fee) from the relevant culture collection, thus facilitating the acquisition of some potentially useful bacteria and fungi. Once acquired, micro-organisms can be maintained and stored as stock cultures using standard microbiological practices (see Appendix A).

Because of the widespread interest in all aspects of microbiology, the numbers and sizes of such culture collections have increased progressively, and the list of such collections maintained world-wide is considerable. Details of some of these open-access culture collections are given in Appendix B. Most culture collections issue catalogues that give basic details for each of the cultures maintained. This information, in accompaniment with a good standard text on microbiology, should identify obligate anaerobes and dangerous pathogens, two categories of micro-organisms that should not be used for biotransformation without appropriate facilities. For reference, a list of the more commonly encountered pathogenic bacteria and fungi is given in Appendix C. Several of the more progressive culture collections file details, such as the name, source, and, most importantly, key biochemical attributes, for all their cultures on computerized records that are, in some cases, accessible to key-word scanning via electronic mail.

Culture collections can also be valuable as sources for additional strains of a useful micro-organism, as well as closely related species, although the cost of buying in these cultures could curtail the scope of this approach. With a basic understanding of the taxonomy of micro-organisms, it is also possible to use such collections to obtain similar micro-organisms from related genera. These approaches can result in some significant improve-

Table 2.5. *Biotransformation of (±)1 to lactones by*
Curvularia lunata NRRL 2380

	Lactones formed (%)			
Fungus	(−)2	(+)2	(−)3	(+)3
Curvularia lunata	33.16	10.84	36.96	19.04
	(75)	(25)	(66)	(34)

Table 2.6. *Biotransformation of (±)1 to lactones by other dematiaceous*
fungi of the genera Curvularia

	Lactones formed (%)			
Fungus	(−)2	(+)2	(−)3	(+)3
Curvularia clavata	41.49	5.26	50.99	2.25
	(89)	(11)	(96)	(4)
Curvularia eragrostridis	41.43	5.20	51.05	2.40
	(89)	(11)	(96)	(4)
Curvularia intermedia	38.15	2.96	57.79	0.86
	(93)	(7)	(99)	(1)
Curvularia leonensii	47.62	15.37	35.27	1.74
	(76)	(24)	(96)	(4)
Curvularia pallescens	43.99	16.86	36.68	1.88
	(72)	(28)	(95)	(5)

ments in the execution of particular biotransformations, as evidenced by
the screening of fungi able to catalyze the enantiodivergent ring expansion
of bicyclo(3.2.0.)hept-2-en-6-one (1) to equivalent isomeric lactones (2)
and (3) (Figure 2.1, Tables 2.5 and 2.6). An initial literature survey identified
Curvularia lunata NRRL 2380 as a potentially useful biocatalyst. Exami-
nation of the culture confirmed that the fungus was able to perform this
oxidative biotransformation, but the enantioselectivity recorded for this
particular fungus was not particularly high. *Curvularia lunata* is a dematiaceous hyphomycetic deuteromycete. A survey of other species of *Curvularia*
and *Drechslera*, a related fungal genus, revealed several micro-organisms
able to undertake an equivalent biotransformation that was more highly
enantioselective. Although it is evidently possible to obtain better microbial
biocatalysts using this approach, success is not guaranteed. For instance, in
a systematic search for micro-organisms able to reduce ketopantoyl

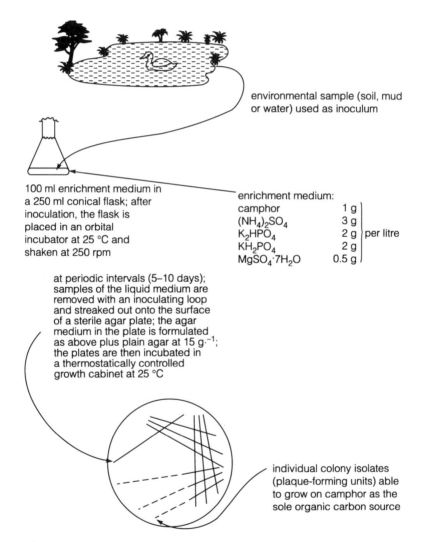

Figure 2.2. Biotransformation of ketopantoyl lactone.

(−)-pantoyl lactone (−)-(5)   ketopantoyl lactone (4)   (+)-pantoyl lactone (+)-(5)

environmental sample (soil, mud or water) used as inoculum

100 ml enrichment medium in a 250 ml conical flask; after inoculation, the flask is placed in an orbital incubator at 25 °C and shaken at 250 rpm

enrichment medium:

camphor	1 g	
$(NH_4)_2SO_4$	3 g	
K_2HPO_4	2 g	per litre
KH_2PO_4	2 g	
$MgSO_4 \cdot 7H_2O$	0.5 g	

at periodic intervals (5–10 days); samples of the liquid medium are removed with an inoculating loop and streaked out onto the surface of a sterile agar plate; the agar medium in the plate is formulated as above plus plain agar at 15 g·$^{-1}$; the plates are then incubated in a thermostatically controlled growth cabinet at 25 °C

individual colony isolates (plaque-forming units) able to grow on camphor as the sole organic carbon source

Figure 2.3. Enrichment selection technique.

lactone (4) to pantoyl lactone (5), one strain of *Rhizopus oryzae* produced predominantly the L-(+)-isomer, whereas another strain of the same fungus produced almost exclusively the D-(−)-isomer (Figure 2.2).

   The picture presented by the type of literature survey described earlier can be useful in another way in obtaining potentially useful micro-organisms. A study of the relevant references listed in Table 2.4 confirms that Baeyer-Villiger mono-oxygenases are essential to enable micro-organisms to grow in media containing cycloalkanones (and cycloalkanols→ cycloalkanones) as the sole source of carbon and energy. Similar micro-organisms can, in principle, be obtained from any natural environment by the so-called enrichment selection technique (Figure 2.3). Typical natural environments that can be sampled are rich soils, which routinely contain a substantial mixed microbial population ($>10^8$ micro-organisms per gram). Such samples are collected and inoculated into a liquid medium containing a cycloalkanone, such as camphor, as the sole source of carbon and energy. Those micro-organisms which are able to metabolize the cycloalkanone will proliferate, unlike those that are unable to use this carbon source. By sub-culturing from such a liquid medium and streaking out the inoculum across the surface of a sterile Petri dish containing an agar formulated with the cycloalkanone as the sole carbon and energy source, it is possible to isolate pure cultures of cycloalkanone-metabolizing micro-organisms. These isolated cultures can then be tested for the ability to undertake oxidative biotransformation of the substrate(s) of interest. Some of the isolates obtained in this way will contain a Baeyer-Villiger mono-oxygenase able to promote the required reaction, thus ensuring access to another source of useful micro-organisms.

2.3.2   What is the most suitable form of the chosen biocatalyst to use?

There are four principal alternative forms of microbial whole-cell bio-catalysts that can be considered. These are

1. growing cultures
2. resting cultures
3. spore cultures
4. immobilized cultures

However, not all of these alternatives are applicable to all micro-organisms; thus the use of spore cultures is restricted to micro-organisms that produce large numbers of easily harvestable spores.

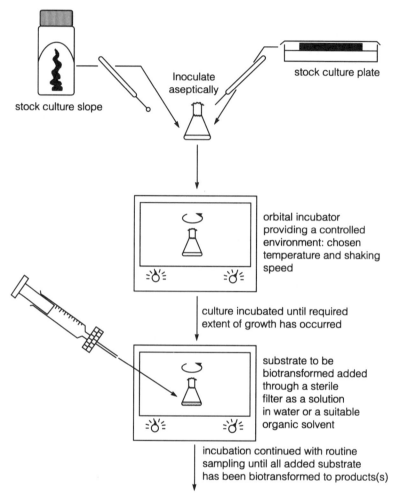

stock culture slope

Inoculate
aseptically

stock culture plate

orbital incubator
providing a controlled
environment: chosen
temperature and shaking
speed

culture incubated until required
extent of growth has occurred

substrate to be
biotransformed added
through a sterile
filter as a solution
in water or a suitable
organic solvent

incubation continued with routine
sampling until all added substrate
has been biotransformed to products(s)

conical flask removed from the incubator; microbial biomass removed by
filtration or centrifugation as appropriate; product(s) extracted from the clear,
spent growth medium

Figure 2.4. Batch culture technique: outline procedure.

Growing cultures

This is the simplest biotransformation procedure; it is based on conventional fermentation practices (see Appendix A). Batch-grown cultures inoculated in conical flasks (50–2,000 ml) are the norm for small-scale bioconversion purposes. In the batch culture technique (Figure 2.4), a pure culture of a micro-organism is grown in a suitable liquid medium. Either at the time of inoculation or at some time thereafter (as discussed later), the substrate to be biotransformed is added to the growth medium, and the incubation is continued. At regular intervals, a fresh sample (<0.5 ml) of the growth medium is removed using a sterile pipette, and the sample is monitored by an appropriate technique (thin layer chromatography, TLC; gas chromatography, GC; high-performance liquid chromatography, HPLC) that is able to record the extent of substrate disappearance and product(s) formation. Incubation is continued until all of the added substrate disappears and/or the bioconversion ceases. In judging the time to stop the fermentation, the level of product(s) accumulating in the growth medium should be scrutinized closely to avoid so-called overmetabolism. Such overmetabolism, which serves to erode the yield of desired product, can occur because cells of the micro-organism may be able to promote enzyme-catalyzed reactions able to undertake one or more further bio-transformation to the product of interest.

Resting cultures

When biotransformations are conducted with growing cultures in growth media, the isolation and purification of the required product(s) can be difficult due to the coincident presence of other metabolites produced by the micro-organisms from components of the medium. The use of a resting cell culture minimizes these problems. Resting cells are non-growing live cells. They are obtained by removing growing cells from an appropriate liquid culture medium at a time in the growth phase when the potential of the cells to undertake the desired biotransformation, as determined by prior experimentation, is optimal. The cells and the substrate to be biotransformed are then resuspended in a weak, compatible buffer solution (20–100 mM) at a pH known to optimize the required biotransformation, which subsequently progresses. The optimum pH of this buffer solution, which has to be determined by prior experimentation, may not be the pH at which the micro-organisms grows best in the culture medium, providing another potential advantage to using resting cultures.

Spore cultures

A spore is a thick walled resistant form of the cell produced by some groups of micro-organisms (notably, most filamentous fungi and some bacteria, including the majority of actinomycetes and species of the genus *Bacillus*) to ride out adverse environmental circumstances, such as elevated temperatures or desiccation. Although widely considered by microbiologists to be metabolically inert, dormant spores are known to undertake a wide range of biotransformations (see the reviews listed in the Bibliography), and thus they can be excellent biocatalysts, because, like resting cells, they can undertake desired reactions when suspended in dilute buffer (or distilled water). Because such spore cultures are incubated in the absence of nutrients, they remain dormant and can continue to promote useful biotransformations over extended periods of time (up to 30 days) without the spores degenerating. Washing the surface of a sporing micro-organism growing aseptically on an appropriate agar in a Petri dish with a dilute solution (0.1% v/v) of a mild detergent such as Tween 80 or Triton X-100 is a convenient way to harvest large numbers of spores, which can either be used immediately (suspended along with substrate to be biotransformed in an appropriate buffer solution) or be stored for extended periods of time (up to 30 months) at $-15\,^{\circ}\text{C}$ until required.

Immobilized cultures

Immobilized cultures of micro-organisms have been much studied and reviewed in recent years (see the Bibliography). All immobilization techniques are based on growing a pure culture of a micro-organism in an appropriate medium until the potential of the cells to undertake a desired activity, as determined by prior experimentation, is optimal. Thereafter, the cells are harvested (filtration or centrifugation, as appropriate), washed to remove the residual growth medium, and then immobilized by an appropriate method based on one of four alternative techniques:

1. Entrapment in a polymer matrix (polyacrylamide, alginate, κ-carrageenan, polyurethane).
2. Surface adsorption onto a water-insoluble solid support [various ion-exchange resins and inorganic polyelectrolytes, e.g. Ti(OH)_4, silicates].
3. Covalent attachment to a water-insoluble solid support [various chemically modified forms of cellulose, e.g. carboxymethylcellulose (CM-cellulose)].
4. Chemical cross-linking with bi-functional agents (e.g. glutaraldehyde, toluene diisocyanate).

Practical experience has confirmed that all of these methods work better with bacteria and yeasts than with filamentous fungi. An illustrative outline protocol is given in Appendix D.

Biotransformations with immobilized microbial cells are attractive in principle for a number of reasons. Provided a suitable immobilization method is used (unless following an established protocol for a particular micro-organism, this will be an empirical choice based on prior experimentation), a immobilized preparation of the cells of a micro-organism is likely to remain operationally active for very much longer than an actively growing culture of the same cells. In addition, an immobilized cell preparation is easily removed from the reaction mixture, thus simplifying work-up of the product(s), and often the reclaimed preparation can subsequently be re-used repeatedly. However, compared to an equivalent volume of growing cells, the apparent catalytic activity of an immobilized cell preparation often is significantly reduced, probably due to limitations on diffusion of the substrate into [and/or product(s) out of] the preparation. In practice this loss of catalytic activity can be compensated by increasing the cell density of the immobilized preparation.

Some of the currently most successful industrial-scale biotransformations exploit such cultures (e.g. the production of L-aspartic acid from fumaric acid by cells of the bacterium *Escherichia coli* immobilized in κ-carrageenan) (see Scheme 6.11 in Chapter 6). Interestingly, there are several reported examples, such as that shown in Figure 2.5 and Table 2.7, where the stereochemical outcome of a biotransformation can be influenced by whether the microbial biocatalyst is used as free cells or immobilized cells; in addition, the nature of the immobilization method itself can be influential in this respect.

2.3.3   *What needs to be done to optimize the relevant enzyme titre(s) of the chosen biocatalyst preparation?*

The overall extent of the biotransformation of a substrate to product(s) by a whole-cell preparation will be determined by the degree to which the reaction is being carried out by each of the individual microbial cells in the culture being used. This will depend on three factors:

1. the number of viable cells per unit volume of the whole-cell preparation,
2. the number of potentially active relevant enzyme molecules per cell,
3. the activity of the relevant enzyme molecules.

Number of viable cells per unit volume

Because of the way that resting cultures, spore cultures, and immobilized cultures are set up for use in biotransformation, the number of cells per unit volume of such preparations can be adjusted by appropriate dilution to any required level. In each case the optimum number of cells per unit volume in such preparations will need to be established empirically for a particular biotransformation.

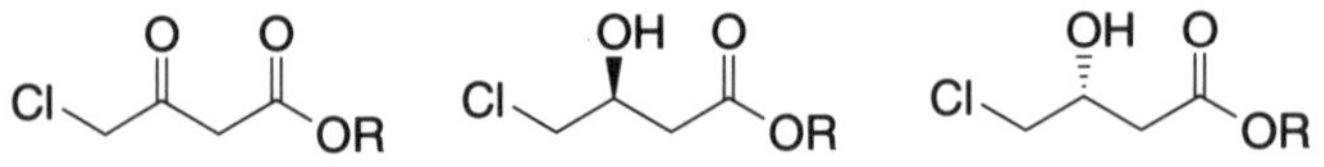

(6) R = Me   (–)-(8) R = Me   (+)-(8) R = Me
(7) R = Et   (–)-(9) R = Et   (+)-(9) R = Et

Figure 2.5. Effects of immobilization on enantiospecificity of β-ketocarboxylic acid reductase of *Saccharomyces cerevisiae*.

Table 2.7. *Effects of immobilization on enantiospecificity of β-ketocarboxylic acid reductase of* Saccharomyces cerevisiae

Substrate and concentration $(mg \cdot ml^{-1})$	Method of immobilization	Product	Enantiomeric excess (%)
(6) (10)	None	(+)-(8)	31
(6) (20)	None	(+)-(8)	12
(6) (20)	Alginate entrapment	(+)-(8)	10
(6) (20)	Polyurethane entrapment	(−)-(8)	90
(7) (10)	None	(+)-(9)	42
(7) (20)	None	(+)-(9)	27
(7) (50)	None	(+)-(9)	15
(7) (20)	Alginate entrapment	(+)-(9)	16
(7) (20)	Polyacrylamide entrapment	(+)-(9)	3
(7) (20)	Polyurethane entrapment	(−)-(9)	82

In the case of growing cultures, the number of viable cells per unit volume of the growth medium is dependent on the age of the culture, the composition of the growth medium (nutritional parameters), and the conditions under which the culture is grown (environmental parameters).

If a stock culture (see Appendix A) of a unicellular micro-organism (e.g. many bacteria and yeasts) is used to inoculate a sterile conical flask containing a medium in which cells of the micro-organism can grow, and the flask is then incubated under conditions that will encourage the culture to grow, the number of viable cells in the flask will vary with time, as shown in Figure 2.6, which represents a typical batch culture growth curve. After inoculation with the stock culture, there is a so-called lag phase, during which there is little or no increase in cell numbers with time. During this lag phase the cells from the inoculum are adjusting to the different growth conditions encountered in the fresh medium, compared with those pertaining in the stock culture. After the lag phase, the cells enter the so-called exponential phase of growth, when the cells begin to grow and divide by binary fission; this is alternatively called the logarithmic (log) phase of growth, because it is a time when the logarithmic number of cells increases linearly with time. Exponential growth continues as long as all nutrients essential for growth are available in the medium and no toxic end-product(s) of metabolism has accumulated to an inhibitory level. However, when the nutrient(s) supply becomes limiting or toxic metabolite(s) accumulation begins, the culture enters the so-called stationary phase, when the increase in the cell number tails off. This is followed some time later by a decline phase, when the lack of nutrient(s) and/or accumulation of toxic metabolite(s) progress to the point where the number of viable cells declines, because many of the cells lyse and consequently die. The durations of the stationary and decline phases will vary substantially, depending on the micro-organism, the medium, and the culture conditions. Note that this traditional method of representing the results is a semi-logarithmic plot; when the data for the lag and log phases are plotted as an arithmetic plot (Figure 2.7), this emphasizes more dramatically the rapid changes in cell numbers that occur during the late exponential phase.

It might be thought that monitoring the increase in total cell weight as a function of time would produce an identical result, but this is not necessarily so, because cells could, as a result of growing in the medium, simply increase the content of storage products, such as polysaccharides or poly-β-hydroxybutyrate.

For filamentous fungi and similar multicellular micro-organisms, the overall form of the growth curve will be similar, because the same considerations apply. However, the actual shape of the curve will be different for such micro-organisms because the multicellular filaments grow by lateral extension, involving division of the cells at the termini.

There are several different ways to measure the changes in cell numbers that occur throughout the growth curve. Despite the foregoing caveat, probably the simplest method of general applicability is to establish by routine sampling how the weight (preferably dry weight) of a unit volume of a homogeneous sample of the culture changes throughout the growth curve. The practicalities of establishing this relationship are easier for micro-organisms that grow uniformly in liquid culture (many bacteria

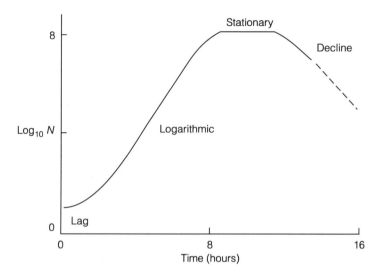

Figure 2.6. Logarithmic plot of a typical batch culture growth curve.

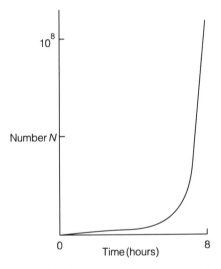

Figure 2.7. Arithmetic plot of a typical batch culture growth curve (same data as used to construct Figure 2.6).

and yeasts) than for those, such as filamentous fungi, that often propagate much less evenly throughout growth media.

The nutritional parameter that directly influences the number of viable cells per unit volume of the growth medium is the availability, as components of the growth medium, of all the nutrients necessary to ensure continuing cell division. The composition of the growth medium can have a profound influence on both the speed and extent of growth, and hence on the form of the growth curve.

If the micro-organism has been obtained from a culture collection or from a researcher in another laboratory, a recommended medium to promote good growth will normally be suggested. If, on the other hand, the micro-organism has been obtained by an enrichment selection technique, then the enrichment medium itself will self-evidently support growth.

If induction effects do not apply, and it is simply lack of growth of a micro-organism that is resulting in a low rate of a desired biotransformation, the nutrient components of the growth medium should be tested for effect one at a time. Both the concentration and type of the carbon source could be influential. Most micro-organisms used for biotransformations are chemo-organoheterotrophs that require an organic molecule as the principal source of carbon and energy. One should first test a range of different carbon sources at one concentration, and then, by a process of elimination, test the apparently better carbon sources at different concentrations. The choice of the carbon sources to try will be dictated both by the type of micro-organism (enrichment isolate/culture collection strain) and by any reported nutritional characteristics. For many micro-organisms able to grow on carbohydrate, growth on polysaccharides, such as starch, is slow because the depolymerizing reaction is rate-limiting; consequently, growth on the equivalent monosaccharide is normally much faster. Carbon sources that are intermediates in the central pathways of biochemistry, such as pyruvate and acetate, are worth trying, although some micro-organisms (e.g. various filamentous fungi) are not able to transport these particular nutrients into cells. Some significant improvements can be recorded, even with micro-organisms supplied with recommended media by culture collections. Thus, whereas the Baeyer-Villiger mono-oxygenase-containing bacterium *Mycobacterium rhodochrous* NCIMB 9784 grows only slowly on the recommended medium containing camphor as the carbon and energy source, much faster growth can be achieved by substituting the chemically related terpene fenchone as the principal carbon source.

Variations in other medium components should be checked, in order of decreasing concentration. When testing nitrogen sources, various complex nitrogen sources such as yeast extract, malt extract, peptones, tryptones, and cornsteep liquor should be tried, as these not only supply readily

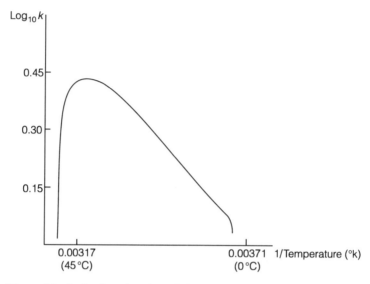

Figure 2.8. Arrhenius plot (log of the growth rate versus temperature^{-1}) for a typical mesophilic micro-organism.

accessible nitrogen, but may additionally satisfy any vitamin and/or trace element requirements. For many micro-organisms, such a nutrient can be used as a combined carbon and nitrogen source.

The environmental parameters most likely to influence growth, and hence the number of viable cells per unit volume of the growth medium, are temperature, pH, and, if applicable, aeration (= oxygen availability). All three parameters can influence both the speed and extent of growth and hence can influence the form of the growth curve.

When using a batch culture technique to grow cells in conical flasks for small-scale biotransformation (see Figure 2.4), the temperature of the incubating cabinet can be used as an important physical variant. Changes in temperature can significantly affect the rates of growth of all micro-organisms. On the basis of this response, micro-organisms are classified into three groups: psychrophiles (temperature optima for growth 0–20 °C), mesophiles (temperature optima for growth 20–50 °C), and thermophiles (temperature optima for growth > 50 °C). Most micro-organisms routinely used for biotransformations are mesophiles. For any micro-organism, an Arrhenius plot (log of the growth rate versus 1/temperature) confirms the expected linear relationship over the appropriate temperature range at which growth can occur (Figure 2.8; plot for a typical mesophilic micro-organism). Clearly there is a temperature optimum above which the growth rate falls off dramatically due to thermal damage to cellular macromolecules.

Similarly, the growth of many micro-organisms is significantly influenced by the pH of the growth medium, with reasonable growth rates being achieved only within about 1–1.5 pH units on each side of an optimum value which is a characteristic of the individual micro-organism. The pH of the growth medium is influenced both by the components used in medium formulation and by the metabolic activities of the micro-organism during growth (nutrients consumed and metabolites accumulated): This is clearly a dynamic relationship. Media are often formulated to contain a buffer system; in addition, a number of commonly used media components are zwitterions. It is reasonably easy to establish empirically for a micro-organism the influence on growth of using any particular medium buffered to different initial pH values.

Most micro-organisms used in whole-cell biotransformations are obligate or facultative aerobes and thus require oxygen for growth when cultured aerobically, which is the only realistic approach for those without access to the specialist facilities required for anaerobic culture. It is difficult to achieve optimum growth in shaken conical flasks, the recommended method of undertaking small-scale whole-cell biotransformations (see Figure 2.4). This is because as a culture of micro-organisms proliferates, oxygen (which is poorly soluble in water) is progressively stripped out of the medium faster than it can be exchanged across the air:liquid interface in the culture flask. Consequently, oxygen becomes a growth-limiting factor. In conical culture flasks, the efficiency of aeration is determined, amongst other things, by the shape of the flask and the volume of liquid medium in it (since both of these factors can affect the size of the air:liquid interface in the culture vessel). The efficiency of aeration is also determined to some extent by the composition of the growth medium (oxygen solubility decreases inversely with solute concentration), the speed of shaking in the incubating cabinet, and the temperature of the incubating cabinet (oxygen solubility decreases inversely with temperature). An indication of the influence of some of these factors is given by the data in Table 2.8.

Some basic rules can help to minimize the growth-limiting effect of oxygen deficit in conical flasks used for small-scale whole-cell biotrans-formations:

1. The liquid volume in a flask should not exceed 20% of the total volume.
2. The shaking rate should be as high as can be safely achieved without

Table 2.8. *Relative oxygen absorption rates in various batch culture flasks*

Glass vessel	Volume of medium (ml)	Incubation conditions	Relative oxygen absorption rate
50-ml conical	10	Stationary	0.13
50-ml conical	20	Stationary	0.04
100-ml conical	20	Stationary	0.16
500-ml conical	100	Stationary	0.05
50-ml conical	10	Shaking, 250 rpm	0.34
50-ml conical	20	Shaking, 250 rpm	0.11
50-ml conical	20	Shaking, 500 rpm	0.19
100-ml conical	20	Shaking, 250 rpm	0.55
500-ml conical	10	Shaking, 250 rpm	1.00
500-ml conical	50	Shaking, 250 rpm	0.30
500-ml conical	100	Shaking, 250 rpm	0.14
500-ml conical	100	Shaking, 500 rpm	0.24

causing splashing in the flasks and/or shear damage to the micro-organisms growing in the medium (as indicated by poor growth rate).
3. The aeration characteristics of a conical flask can be increased substantially by attaching baffles to the inside surface of the base of the flask (Figure 2.9).

Ignoring, for the moment, additional considerations arising from any variation in the number of relevant enzyme molecules per cell, there are various factors to consider when contemplating the optimum time for the addition of a substrate to be biotransformed. Normally, the ideal time to add a substrate for bioconversion is during the middle or late exponential phase, thus ensuring the growing cells exposure to the substrate over the time window when the number of viable cells in the culture will be maximal. Adding the substrate during the stationary or decline phase normally results in low rates of bioconversion, except in some special circumstances, such as when the relevant enzyme(s) is present exclusively during so-called secondary metabolism. This recommended time course will coincidentally minimize any potential toxicity problems resulting from addition of the substrate. Toxic substrates can result in low rates of biotransformation if added to the growth medium during the lag phase or early exponential phase of growth.

Number of potentially active relevant enzyme molecules per cell

The enzyme complements for growing cultures can be divided into two categories: constitutive and adaptive enzymes. Constitutive enzymes are present to the same (or very similar) extent in microbial cells throughout the growth curve. Such enzymes have the same (or similar) copy number (i.e. extent of transcription from relevant genes) throughout each of the phases of active growth. If, after suitable experimentation, it is apparent that a biotransformation of interest occurs throughout all stages of growth on various different media, then it is likely that this is due to the activity of such a constitutive enzyme(s). In these circumstances the optimum time for addition of the substrate to be biotransformed is normally the middle exponential phase, based on considerations of the numbers of viable cells, as discussed earlier.

Adaptive enzymes, however, are present to different extents at different times throughout the growth curve. For the purposes of promoting various biotransformations, an important group of adaptive enzymes is composed of the inducible enzymes. The characteristic feature of inducible enzymes is that they are not normally synthesized by cells unless a relevant inducer molecule is present in the cells to promote selective expression of the equivalent genomic information. The well-known models proposed by Jacob and Monod for prokaryotic micro-organisms (bacteria) and by Britten and Davidson for eukaryotic micro-organisms (includes yeasts and fungi) offer elegant explanations of how the copy numbers of such inducible enzymes can vary in response to the presence in cells of appropriate inducers. Inducers are normally highly specific in their modes of action and control the expression of one (monocistronic induction) or, in some cases, a small number of biochemically interactive adaptive enzymes (polycistronic induction, co-ordinate induction). Two different types of inducible enzyme responses in micro-organisms are recognized, both of which are significant when undertaking whole-cell biotransformations with micro-organisms exhibiting these phenomena.

In some cases, the adaptive enzyme response is clearly intended to enable micro-organisms to grow on particular nutrient resources when available as components of the growth medium. A characteristic feature of such an enzyme is that high titres are present in cells only during growth on media formulated to contain the relevant exogenous nutrient inducer. A well-known example of such an inducer is the disaccharide lactose, which, as recognized by Jacob and Monod, is responsible for inducing the hydrolytic enzyme β-galactosidase in the bacterium *Escherichia coli*, thus enabling cells of the micro-organism to grow on a medium containing lactose as the sole source of carbon and energy. If a biotransformation of interest is being catalysed by such an inducible enzyme, then suitable experimentation must be undertaken to determine the following:

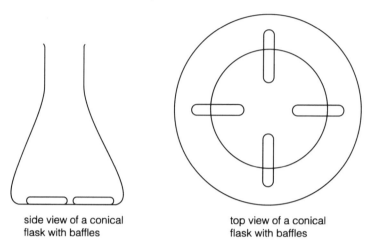

side view of a conical
flask with baffles

top view of a conical
flask with baffles

Figure 2.9. Conical flask with baffles to enhance aeration of shaken liquid contents.

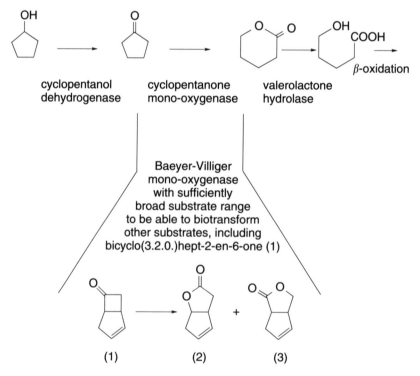

cyclopentanol
dehydrogenase

cyclopentanone
mono-oxygenase

valerolactone
hydrolase

β-oxidation

Baeyer-Villiger
mono-oxygenase
with sufficiently
broad substrate range
to be able to biotransform
other substrates, including
bicyclo(3.2.0.)hept-2-en-6-one (1)

(1)   (2)   (3)

Figure 2.10. Cyclopentanol-induced enzymes involved in alicyclic ring cleavage in
Pseudomonas sp. NCIMB 9872.

1. What is a suitable inducer for the enzyme of interest, and consequently what further additions, if any, need to be made to the growth medium to enhance the enzyme titre of the cells? Inducible enzymes, like all enzymes, are participants in one or more of the biochemical pathways that function in the cells of micro-organisms. For most inducible enzymes, a suitable inducer will be either the substrate for the enzyme itself or, if co-ordinately induced, the substrate for an enzyme that occurs earlier in the pathway in which the enzyme of interest participates. Thus, in the case of the biotransformationally useful Baeyer-Villiger mono-oxygenase in *Pseudomonas* sp. NCIMB 9872 (Figure 2.10), the best inducer for this enzyme is the growth substrate cyclopentanol, because the three initial enzymes involved in cycloalkanol catabolism in this bacterium (cyclopentanol dehydrogenase, cyclopentanone mono-oxygenase, and valerolactone hydrolase) are all coordinately induced by the alicyclic alcohol in the growth medium. A basic understanding of the biochemical pathways that function in microorganisms is undeniably helpful in this aspect of biotransformations.

2. How does the titre of the inducible enzyme change throughout the growth curve? Knowing how the enzyme activity changes during the various stages of growth, allied with considerations based on the number of cells in the culture during the different growth phases, helps to establish the optimum time for the addition to the growth medium of the substrate to be biotransformed. As illustrated by the example of the Baeyer-Villiger mono-oxygenase in cyclopentanol-grown *Pseudomonas* sp. NCIMB 9872 (Figure 2.11), such changes can be very substantial. In this particular case the optimum time to add to the growth medium the ketones to be biotransformed, such as bicyclo-(3.2.0.)hept-2-en-6-one, is during the late exponential phase. Ketone addition either earlier in the middle of the log phase or later in the early stationary phase will result in relatively poor yields of oxygenated biotransformation products.

In other cases, the adaptive enzyme response is not involved with initiating growth on particular nutrients, but is instrumental in triggering some form of developmental change in the cells of micro-organisms, such as the formation of spores, which is a characteristic of some bacteria (e.g. *Bacillus* species) and very many fungi. These profound adaptive biochemical changes that result in sporulation are associated with the widely recognized phenomenon of the "swap" from so-called primary metabolism (growth-oriented metabolism) to secondary metabolism (post-

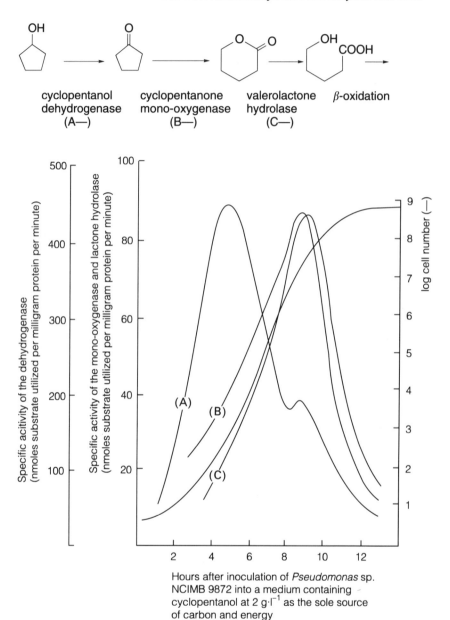

Figure 2.11. Growth of *Pseudomonas* sp. NCIMB 9872 on cyclopentanol.

growth metabolism). Such patterns of developmentally programmed adaptive change are often poorly characterized, but are usually triggered by endogenous inducers produced within the microbial cells themselves. As such, this type of adaptive change normally cannot be influenced by adding nutrients as putative inducers to the growth medium. This phenomenon is not widely appreciated by researchers in the field of biotransformations, possibly because of limited knowledge of the attributes of micro-organisms. However, media formulation is not entirely divorced from influencing such developmental events, because it is widely accepted that growth of micro-organisms in nutritionally unbalanced media can trigger sporulation. The balance of available carbon:nitrogen nutrients in media is often the critical parameter in this respect. Some relatively simple experimentation in changing the concentrations and types of carbon and/or nitrogen source in media often can result in such phenomena being predictably managed in batch culture. For instance, the useful Baeyer-Villiger mono-oxygenase enzyme that occurs in dematiaceous fungi such as *Curvularia lunata* NRRL 2380 is known to be endogenously induced as a part of the package of developmental changes, including the production of copious amounts of melanin-based pigments, associated with the swap from primary to secondary metabolism and the concomitant onset of sporulation. By growing *C. lunata* NRRL 2380 in a medium formulated with glucose at $10 \, \text{g} \cdot \text{l}^{-1}$ and cornsteep liquor at $20 \, \text{g} \cdot \text{l}^{-1}$, and adding ketone substrates to be biotransformed [e.g. bicyclo(3.2.0.)hept-2-en-6-one] at the time that the fungal mycelium begins to darken, 60 hours after inoculation, one can consistently produce high levels of synthetically useful lactones. However, if the initial glucose content of the medium is increased to $40 \, \text{g} \cdot \text{l}^{-1}$, the onset of sporulation is considerably delayed (> 300 hours) and becomes much less predictable, which can result in the optimum time for ketone substrate addition being missed.

When using microbial biocatalysts other than actively growing cells, both resting cultures and immobilized cultures should be prepared from growing cells harvested at an appropriate point in the growth curve. Spores have a species-specific enzyme complement that, being developmentally programmed, is invariable and is not amenable to adaptive control. One considerable advantage of spores is that subsequent to formation, and prior to germination, they exhibit no apparent transcriptional activity. Consequently, biotransformations that exploit those enzymes that are active in spore preparations are characteristically consistent and can be continued in a predictable manner over extended periods of time.

Activity of enzymes

Temperature and pH are two environmental parameters that are known to affect the catalytic activity of isolated enzymes and thus are potentially influential in whole-cell biotransformations. When using growing cultures, any effects of changes in the pH of the culture medium or the incubation temperature on the activities of individual enzymes are likely to be direct reflections of the effects of these environmental parameters on the growth of the cells.

The active enzyme complement in most spore cultures is singularly unresponsive to changes in pH and temperature. This probably is a reflection of the fact that spores were evolved to be thick-walled cell forms, highly resistant to considerable extremes of environmental circumstances.

There is considerably more scope for using temperature or pH to influence the activities of target enzymes when one is employing resting and immobilized cultures, since the optimum for biotransformation can well be different to the optimum for growth. When using such preparations to undertake a particular biotransformation, a relatively simple programme of experiments can be undertaken to establish the optimum value for any environmental parameters, such as pH or temperature. Resting cultures of a wide range of micro-organisms frequently perform well at temperatures considerably in excess of the temperatures for optimal growth. Similarly, for immobilized cultures, the effects of pH often are far less restrictive than are observed with the equivalent growing cells, possibly due to the effects of diffusional limitation imposed by the immobilization matrix.

2.3.4 What, if anything, needs to be done to optimize biocatalyst–substrate interaction?

An organic compound is amenable to biotransformation only if it is accessible to the relevant enzyme complement of a whole-cell biocatalyst. This presents little or no problem for those biotransformations catalysed by enzymes secreted outside the boundary wall by the cells of micro-organisms. However, for the majority of whole-cell biotransformations, the relevant enzymes are retained within the cells of the micro-organisms.

Compounds for biotransformation that are water-soluble, with ionizable and/or hydrophilic functional groups, are relatively easy to handle and can be added directly to the various types of whole-cell preparations, since they are located in aqueous-based environments. Some problems may arise above a threshold concentration with particular hydrophilic compounds, however, resulting either from the toxicity to some micro-organisms of certain compounds or from ionic and/or pH changes that accompany such additions. Both spore and immobilized cultures tend to be relatively immune to such effects. A simple programme of experimentation can identify and surmount these particular problems. Detrimental excessive ionic changes can often be countered by increasing the buffering capacity of the medium containing the cell preparation, although high solute levels may exacerbate an oxygen supply problem for growing cultures. Substrates that are toxic can be added incrementally throughout the period when the relevant enzyme titre of the cells is maximal. Typically, it was found that when added as a single aliquot during the appropriate phase of the growth curve, the highest level of 7,7-dimethylbicyclo(3.2.0.)hept-2-en-6-one (10) that could be biotransformed into equivalent lactones (11) and (12) (Scheme 2.2) by a growing culture of the Baeyer-Villiger mono-oxygenase-containing fungus *Mortierella isabellina* NRRL 1757 was $1.5 \, g \cdot l^{-1}$. However, when the ketone was sequentially added as a series of $0.2 \, g \cdot l^{-1}$ aliquots throughout the active phase of the growth curve, the fungus could be made to biotransform a total addition of the substrate of $4.2 \, g \cdot l^{-1}$. This stratagem can have an added benefit in avoiding excess substrate inhibition, a kinetic characteristic of some enzymes which occurs when the relevant substrates are used above a critical concentration. In some cases, incremental substrate feeding additionally increases product yield by minimizing substrate loss via undesirable competing reactions: this probably reflects differing kinetic characteristics of the competing enzyme systems.

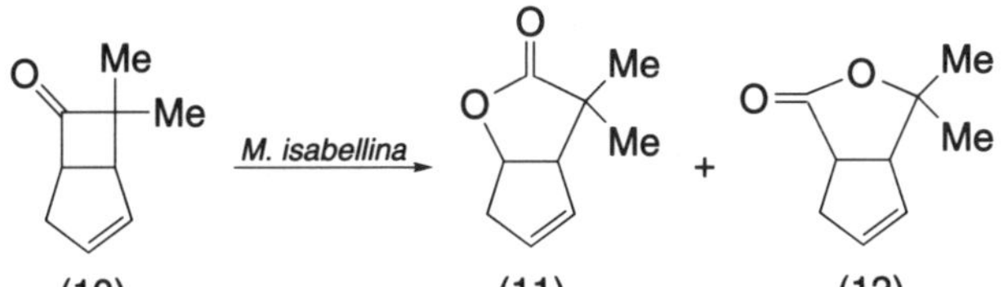

(10)   (11)   (12)

Scheme 2.2

Hydrophobic compounds for biotransformation may pose similar problems, which can also be dealt with by incremental feeding. However, the major problem with such substrates often is accessibility to the bio-catalyst. The most common method for adding hydrophobic substrates is as a solution in the minimum possible volume of a water-miscible organic solvent. Such organic solvents preferably should have low toxicity to the chosen microbial biocatalyst. This is particularly important when using growing cultures, of less importance for resting cells, and of relatively little significance for spores. The solvents that most frequently prove successful include *N*,*N*-dimethylformamide (DMF), dimethyl sulphoxide, ethanol and acetone. Of these, DMF is particularly useful because it is a suitable solvent for a wide range of lipophilic substrates, and it is apparently less toxic than the other solvents to a wide range of different micro-organisms. Although it is possible in some cases to use surfactants such as Tween 80, Triton X-100 and Nonidet P40 as alternative delivery systems (rather than organic solvents) for adding hydrophobic substrates, such additions usually are less effective for enhancing equivalent biotransformations. However, the combined solubilizing effects of both an organic solvent and a detergent often are greater than those for either treatment used individually. As well as promoting adequate dispersion of hydrophobic substrates in aqueous environments, both organic solvents and detergents serve to increase the permeability of microbial cells by influencing the structural integrity of the membranes present in the cells. Clearly, the permeability of the peripheral plasma membrane is very influential in this respect.

Other techniques that have proven successful in presenting some hydrophobic substrates to whole-cell preparations include sonication, adsorption onto polyelectrolytes, and, in the case of some solid substrates, fine milling. Sonication is a well-established technique able to produce micro-emulsions that can significantly aid substrate dispersion in aqueous environments. Sonication, like organic solvents and surfactants, can also be used to permeabilize resting cultures, although some simple experimentation is necessary to establish a compromise protocol that will aid substrate uptake but not jeopardize the viability of the cells beyond an acceptable extent. A variety of hydrophobic compounds can be adsorbed onto both natural (silicates, zeolites) and synthetic (Dowex, Amberlite, Duolite) inert polyelectrolyte matrices. Because of the ultrafine particle sizes of these materials, the resultant large surface area for the adsorbed material ensures a high degree of dispersion when added to aqueous suspensions of whole-cell microbial biocatalysts. Milling is a technique originally developed for steroids substrates, and it has been found useful for various other solid hydrophobic substrates. The substrate is ground to fine particle size using an appropriate mechanical device.

2.3.5 *What, if anything, needs to be done to optimize product yield by preventing alternative metabolic fates for the substrate and/or product?*

The major problem here results from the use of whole-cell biocatalysts, which in effect represent a "package deal" of several hundreds of different

types of enzymes, one or more of which undertake the desired biotransformation, but one or more of which may, alternatively, provide either of
the following:

1. An additional fate other than that sought for the added substrate. Such
   undesirable reactions decrease the overall efficiency of the required
   reaction, by competing for added substrate, and may considerably
   complicate the subsequent procedures required to purify the sought
   products from the spent reaction mixture. In some cases it may prove
   possible, by appropriate experimentation, to minimize such problems
   kinetically by determining a level of substrate addition that will favour
   the sought reaction. However, in other cases an alternative solution to
   this problem needs to be found.
2. A route whereby the sought product(s) is further biotransformed.
   This phenomenon, referred to as "overmetabolism", also results in a
   decrease in the overall efficiency of the required reaction.

An example illustrating both problems, and some suitable solutions, is
provided by the Baeyer-Villiger mono-oxygenase-dependent biotransformation of bicyclo(3.2.0.)hept-2-en-6-one to synthetically useful lactones
by a resting culture of the cyclopentanol-grown bacterium *Pseudomonas*
sp. NCIMB 9872 (Figure 2.10).

As well as a Baeyer-Villiger mono-oxygenase able to undertake the required oxidative biotransformation of added bicyclic ketone, cyclopentanol-
grown cells used to produce a resting culture optimal for lactone formation
also contain significant levels of a second type of mono-oxygenase, able
to promote the regiospecific hydroxylation of the bicyclic substrate to the
4'-*exo*-hydroxyketone. This serves to decrease the yield of the sought
lactones by providing an alternative sink for some of the added substrate.
The hydroxylase responsible, like a number of similar microbial enzymes,
is an iron-containing protein that can be completely inhibited by adding
to the resting cultures various metal-chelating agents, including α,α'-
dipyridyl (0.0005 M), *o*-phenanthroline (0.0001 M) and 8-hydroxyquinoline
(0.0005 M). The value of these and similar metal chelators (Table 2.9) in
preventing unrequired reactions was first realized during studies to
promote the accumulation of certain required products of steroid biotransformation. Such inhibitors work well with resting cultures. Although there
are some notable exceptions, they are generally less successful when used
with growing cultures, probably because of their non-specific mode of
action. The effects of such inhibitors with spore and immobilized cultures
have not been extensively studied.

Table 2.9. *Some useful enzyme inhibitors, classed by mode of action*

Reagents that inhibit metalloenzymes
1. *Metal-ion-chelating agents*
   ammonium purpurate
   α,α'-dipyridyl
   desferrioxamine **B**
   diethylenetriaminopentaacetic acid
   2,3-dimercapto-1-propanol
   4,5-dihydroxybenzene-1,3-disulphonic acid
   ethylenediaminetetraacetic acid (EDTA)
   (ethylenedioxy)diethylene nitrilotetraacetic acid (EGTA)
   o-hydroxybenzaldehyde oxime
   nitrilotriacetic acid
   N,N-bis(hydroxyethyl)glycine
   o-phenanthroline (1,10-phenanthroline)
   salicylic caid
   sodium diethyldithiocarbamate

2. *Other modes of action*
   sodium azide
   sodium cyanide
   sodium fluoride

Reagents that inhibit flavoproteins
atebrin hydrochloride (quinacrine)
quinine
sodium 5-ethyl-5-isoamylbarbiturate

Reagents that inhibit pyridoxal phosphate–dependent enzymes
o-(carboxymethyl)hydroxylamine

Reagents that inhibit thio-containing proteins
1. *Alkylating reagents*
   iodoacetamide
   sodium iodoacetate
2. *Mercaptide-forming reagents*
   p-hydroxymercuribenzoate
3. *Oxidizing agents that convert thiols to S–S*
   iodosobenzoate
4. *Others*
   N-ethylmaleimide

Organophosphorus reagents that inhibit hydrolases (serine hydrolases)
diethyl *p*-nitrophenyl phosphate (paraoxon)
diisopropyl phosphofluoridate (DFP)
dimethyl(dichlorovinyl) phosphate (dichlovos)
O,S-methylphosphoramidothioate
3-hydroxy-*N*-methyl-*cis*-crotonamide dimethylphosphate
tetraethyl pyrophosphate (TEPP)

There are numerous other types of irreversible enzyme inhibitors, other than those that act as chelating agents for metal ions, that can be considered as putative antagonists for unrequired reactions. Illustrative examples of irreversible enzyme inhibitors belonging to a number of these additional categories are given in Table 2.9. Although there are some general guidelines that may help to indicate in particular circumstances the likely degree of success when using certain enzyme inhibitors (e.g. organophosphorus compounds are highly active against many types of hydrolases), their potential roles in preventing alternative unrequired fates for particular added substrates normally must be determined empirically by appropriate experimentation.

Irreversible enzyme inhibitors also provide a way to tackle the problem of overmetabolism that results in low yields of desired product(s). One such illustrative example of desired product attrition is again provided by *Pseudomonas* sp. NCIMB 9872. Resting cultures of cyclopentanol-grown cells prepared to contain the optimal active titre of Baeyer-Villiger mono-oxygenase also contain high levels of the lactone hydrolase, which metabolizes valerolactone to 5-hydroxyglutarate (Figure 2.11). This enzyme, because it is also able to yield equivalent hydroxycarboxylic acids from the synthetically useful lactones formed from bicyclo(3.2.0.)hept-2-en-6-one, serves to overmetabolize these sought biotransformation products. In this case, probably because of the co-ordinate induction of the relevant genes by the growth substrate cyclopentanol, it is not feasible to produce a whole-cell preparation with a high titre of the Baeyer-Villiger mono-oxygenase but a complementary low titre of the lactone hydrolase. However, it is possible to prevent this attrition of the lactones, and thereby increase the overall accumulation of useful lactones, by including low levels (0.0005 M) of hydrolase inhibitors such as tetraethyl pyrophosphate (TEPP) or dimethyl(2,2-dichlorovinyl) phosphate (dichlorvos) in the reaction mixture. This stratagem is feasible because the biocatalyst is a resting culture of the cyclopentanol-grown bacterium and so consists of cells that are no longer dependent on an active lactone hydrolase to achieve continued growth at the expense of valerolactone. Clearly the same approach would not be appropriate with a growing culture of the same biocatalyst.

As an alternative to the use of metabolic inhibitors to prevent an unrequired reaction, it is possible in some circumstances to consider the use of a mutant strain appropriately modified to eliminate the problematical enzyme-catalysed step(s). Some of the basic techniques of mutagenesis involved in using either physical (ultraviolet light) or chemical (e.g. nitrosoguanidine, ethyl methanesulphonate) mutagens to produce random changes in the genomic complement of a micro-organism are relatively simple to execute. However, the key to success in such an approach is to devise and implement suitable selection protocols that will identify the few survivors characterized by a relevant metabolic lesion (blocked mutants) from amongst the larger number of viable cells remaining after implementation of a random mutation programme. Operating even the proven selection screens to obtain mutant survivors that are unable to undertake unrequired reactions is not recommended, without guidance, for anyone who does not have considerable knowledge of the biochemical pathways that function in micro-organisms and experience of the relevant methodology. Some reviews covering this aspect of whole-cell biotransformations using micro-organisms are listed in the Bibliography.

2.4 Conclusions and overview

Biotransformations of organic compounds using whole-cell microbial biocatalysts involve a multidisciplinary approach that combines knowledge and skills from biochemistry, microbiology and organic chemistry. Indeed, close collaboration between a microbial biochemist and an organic chemist can be most productive in terms of the science that emerges. Nevertheless, it should be emphasized that non-specialist bench chemists can learn the practices involved in growing and using many microbial cells in a matter of a few weeks (obviously the use of baker's and brewer's yeasts is the most facile point of entry to this aspect of the science). Acquiring and assimilating the background knowledge takes considerably longer!

There are many different ways to exploit enzymes and micro-organisms so as to undertake a diverse range of biotransformations. By having the capability of carrying out isolated-enzyme and whole-cell bioconversions, the whole of this range is available to the experimentalist. Such bioconversions can be used to prepare new intermediates for the synthesis of important and valuable fine chemicals. These aspects will be dealt with in the next chapters. At this time (February, 1994), research and development work is split approximately equally into three categories:

1. hydrolysis reactions
2. oxidation and reduction reactions
3. reactions involving carbon–carbon, carbon–nitrogen, carbon–chalcogen and carbon–oxygen bond formation, with particular emphasis on carbohydrate chemistry

Reactions in category 1 (Chapter 3) and category 3 (Chapter 5) are generally carried out using isolated enzymes. Reactions in category 2 (Chapter 4) are most often accomplished using whole-cell systems.

Appendix A

Growing and maintaining micro-organisms

General equipment
Erlenmeyer flasks (100–2,000 ml)
Balance
Beakers
Bunsen burner
Inoculating loop
Inoculating needle
Measuring cylinders
pH meter
Autoclave (alternatively, for relatively small-scale biotransformations it is possible to use a domestic pressure cooker)
Thermostatically controlled orbital incubator (alternatively, it is possible to use either a rotary-action shaker in a temperature-controlled room or a water-bath shaker)

General Procedures
Most micro-organisms will grow either in liquid media (broths) or on the surfaces of solid media (agars) that contain all the nutrients necessary to encourage cell proliferation. Agars are most usually used to maintain stock cultures of micro-organisms, whereas broths are normally used for growing large numbers of cells for whole-cell biotransformations. The most convenient forms of stock cultures are obtained by growing micro-organisms on agar plates. These are prepared prior to inoculation of the desired stock culture using a suitable growth-promoting agar dispensed in sterile plastic Petri dishes that can be obtained from various specialist suppliers.

When growing micro-organisms either on agars or in broths, the protocol used must incorporate the following general features:

1. The medium must initially contain no micro-organism; such a medium is said to be "sterile". The preparation of a sterile medium normally involves the use of an autoclave.

2. Sterile technique is used to ensure that the sterile medium is inoculated in such a way that only the desired micro-organism is introduced and encouraged to grow, with other micro-organisms – contaminants – being excluded. A medium inoculated and subsequently cultured in this way is said to be "aseptic".

3. The inoculated medium must be incubated at a suitable temperature to encourage the micro-organisms to grow.

Preparing media

Media are prepared to defined formulations using requisite amounts of appropriate chemicals. Many of the more commonly used media, such as nutrient broth, tryptone soya broth, malt agar, and potato dextrose agar, are available in pre-mixed dehydrated form from various specialist suppliers. Any components of media that are heat-sensitive must be added to the remainder of the medium after it has been heat-sterilized by autoclaving; this is normally done by syringing in these particular components as concentrated aqueous solutions through disposable sterile filter units available from various specialist suppliers.

Broths and agars should be prepared in appropriately sized Erlenmeyer flasks using distilled water, unless otherwise specified. When preparing broths in which micro-organisms will subsequently be grown, a useful general guideline is that the volume of the contents of the flask should be no more than about a fifth of the total volume of the flask. This not only avoids splashing of the medium contents of the flasks when they are subsequently shaken during culture of the micro-organisms, but more importantly it ensures adequate aeration of the growing culture from the head-space above the medium in the flasks (see Table 2.8). For the preparation of solid media, the agar gelling agent is included at 1.2–1.5% w/v. The required chemicals are dissolved in distilled water by stirring and gentle heating, if required, although care is necessary to avoid charring when preparing agars. The pH of the medium is then adjusted, if necessary, using either 0.1-M NaOH or HCl. The flasks are then plugged with non-absorbent cotton-wool and capped with aluminium foil prior to being sterilized in an autoclave.

Flasks of broth or agar, complete except for any heat-sensitive components of the medium formulation, are sterilized in an autoclave under appropriate conditions (usually steam pressure of 15 pounds per square inch for 15 minutes). For small-scale biotransformations it is possible to use a domestic pressure cooker for steam sterilization.

After sterilization, flasks of broth are allowed to cool prior to using sterile technique to inoculate the flasks with the chosen micro-organism. For bacteria, a sterile inoculating loop is used; for fungi, it is often easier to use an inoculating needle to remove a piece of agar carrying a pre-grown culture of the fungus from a stock plate. Inoculated flasks are then incubated under conditions favourable for growth of the micro-organism to yield a whole-cell culture suitable for undertaking biotransformations.

The properties of agar powder (that it will dissolve readily when autoclaved and remain in a molten state at about 50 °C) make it ideal for preparing sterile agar plates on which to propagate stock cultures of micro-organisms. After autoclaving, flasks of sterile agar are cooled to just above the set temperature by being placed in an incubator or water bath at 50 °C for 60 minutes. The sterile agar is then dispensed into a series of sterile plastic Petri dishes (available from various suppliers) so that the base of each Petri dish is covered about 5 mm deep. After the agar has solidified in the plates (approximately 5 minutes), sterile technique is used to inoculate the surface of the agar with a micro-organism of interest. As when inoculating broth cultures, it is advisable to use an inoculating loop to transfer bacteria, but an inoculating needle for fungi. Inoculated plates are then incubated at a temperature favourable for growth of the micro-organisms, the resulting surface growth representing a stock culture.

Appendix B

Details of some useful culture collections

ATCC	American Type Culture Collection 12301 Parklawn Drive Rockville, Maryland 20852, USA
CBS	Centraalbureau voor Schimmelcultures Julianalaan 67, 2628 BC Delft, The Netherlands
IMI (CMI)	International (Commonwealth) Mycological Institute Ferry Lane Kew, Surrey, TW9 3AF, England
DSM	Deutsche Sammlung von Mikroorganismen und Zellkulturen Mascheroder Weg 1b D-3300 Braunschweig, Germany

IFO	Institute of Fermentation Osaka 17-85 Jusohomachi 2-chome Yodogawaku, Osaka 532, Japan
NCIMB	National Collections of Industrial and Marine Bacteria 23 St. Machar Drive Aberdeen AB2 1RY, Scotland
NCYC	National Collection of Yeast Cultures Food Research Institute Colney Lane, Norwich Norfolk NR4 7UA, England
NRRL	Northern Regional Research Laboratory Agricultural Research Service, USDA 1815 N. University St. Peoria, Illinois 61604, USA
VKM	All-Union Collection of Microorganisms Department of Culture Collection Institute of Biochemistry and Physiology of Microorganisms Academy of Sciences Pushchino, Moscow Region 142292, Russia

Appendix C

Pathogenic bacteria and fungi

Dangerous Pathogens

BACTERIA

Bacillus anthracis	*Mycobacterium bovis*
Bordetella pertussis	*Mycobacterium leprae*
Clostridium bifermentans	*Mycobacterium tuberculosis*
Clostridium botulinum	*Neisseria gonorrhoeae*
Clostridium fallax	*Neisseria meningitidis*
Clostridium histolyticum	*Pasteurella pestis*
Clostridium oedematiens	*Pseudomonas pseudomallei*
Clostridium septicum	*Salmonella typhi*
Clostridium welchii (perfringens)	*Streptococcus pneumoniae*
Corynebacterium diphtheriae	*Treponema pallidum*
Flavobacterium meningosepticum	*Treponema pertenue*
Leptospira icterohaemorrhagiae	*Vibrio cholerae*

FUNGI

Aspergillus fumigatus
Blastomyces dermatitidis
Coccidioides immitis

Histoplasma capsulatum
Histoplasma farcimimosum
Paracoccidioides braziliensis

Less Dangerous but Readily Infectious Pathogens

BACTERIA

Borrelia species
Brucella abortus
Brucella melitensis
Brucella suis
Fusobacterium fusiforme
Haemophilus aegyptius
Haemophilus ducreyi
Haemophilus influenzae
Klebsiella pneumoniae
Klebsiella rhinoscleromatis

Moraxella lacunata
Pasteurella tularensis
Pseudomonas aeruginosa
Pseudomonas pyocyanae
Shigella dysenteriae
Shigella flexneri
Shigella sonnei
Staphylococcus aureus
Streptococcus pyogenes

FUNGI

Candida albicans
Epidermophyton flossosum
Microsporum species

Sporotrichum schenkii
Trichophyton verrucosum

Appendix D

A typical immobilization protocol: entrapment in κ-carrageenan

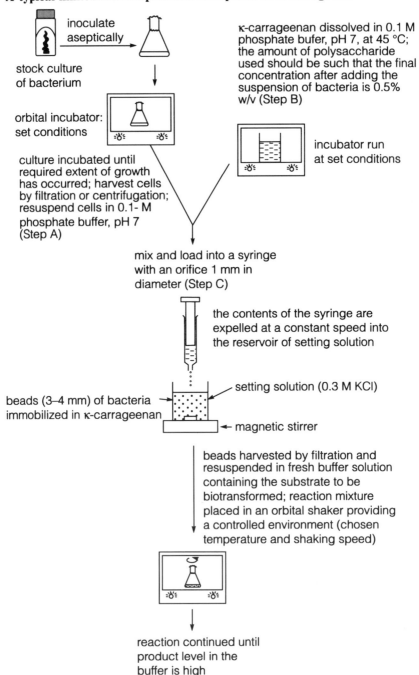

stock culture of bacterium

inoculate aseptically

κ-carrageenan dissolved in 0.1 M phosphate bufer, pH 7, at 45 °C; the amount of polysaccharide used should be such that the final concentration after adding the suspension of bacteria is 0.5% w/v (Step B)

orbital incubator: set conditions

culture incubated until required extent of growth has occurred; harvest cells by filtration or centrifugation; resuspend cells in 0.1- M phosphate buffer, pH 7 (Step A)

incubator run at set conditions

mix and load into a syringe with an orifice 1 mm in diameter (Step C)

the contents of the syringe are expelled at a constant speed into the reservoir of setting solution

setting solution (0.3 M KCl)

beads (3–4 mm) of bacteria immobilized in κ-carrageenan

magnetic stirrer

beads harvested by filtration and resuspended in fresh buffer solution containing the substrate to be biotransformed; reaction mixture placed in an orbital shaker providing a controlled environment (chosen temperature and shaking speed)

reaction continued until product level in the buffer is high

Bibliography

Books and reviews detailing micro-organisms that biotransform particular types of substrates

Steroids

Charney, W., & Herzog, W. (1967). *Microbial Transformations of Steroids. A Handbook*, 2nd ed. Academic Press, New York.
Iizuka, H., & Naito, A. (1981). *Microbial Transformation of Steroids and Alkaloids*. University of Tokyo Press.
Kieslich, K. (1980). Steroid Conversions. In *Economic Microbiology*, vol. 5, ed. A. H. Rose, pp. 369–465. Academic Press, London.
Miller, T. L. (1985). Steroid Fermentations. In *Comprehensive Biotechnology*, vol. 3, ed M. Moo-Young, pp. 297–318. Pergamon Press, Oxford.

Non-steroid cyclic compounds

Kieslich, K. (1976). *Microbial Transformations of Non-steroid Cyclic Compounds*. Georg Thieme Verlag, Stuttgart.

Antibiotics

Sebek, O. K. (1980). Microbial Transformations of Antibiotics. In *Economic Microbiology*, vol. 5, ed. A. H. Rose, pp. 575–612. Academic Press, London.

Alkaloids and nitrogenous xenobiotics

Vining, L. C. (1980). Conversion of Alkaloids and Nitrogenous Xenobiotics. In *Economic Microbiology*, vol. 5, ed. A. H. Rose, pp. 523–73. Academic Press, London.

Various (Antibiotics, hydrocarbons, prostaglandins, xenobiotics, alkaloids, cannabinoids, pesticides)

Rosazza, J. P. (ed.) (1982). *Microbial Transformation of Bioactive Compounds*, 2 vols. CRC Press, Boca Raton, Fla.

Various (Steroids and non-steroid alicyclic hydrocarbons)

Fonken, G. S., & Johnson, R. A. (1972). *Chemical Oxidations with Microorganisms*. Marcel Dekker, New York.

Various (Steroids, sterols, terpenoids, alicyclic and heterocyclic hydrocarbons, alkaloids, antibiotics, aromatic hydrocarbons, aliphatic hydrocarbons, amino acids and peptides, carbohydrates)

Rehm, H.-J., & Reed, G. (eds.) (1984). *Biotechnology, Vol. 6a, Biotransformations*. Verlag Chemie, Weinheim.

Books and reviews detailing the use of spore cultures

Vezina, C., Sehgal, S. N., & Singh, K. (1968). Transformation of Organic
    Compounds by Fungal Spores, *Adv. Appl. Microbiol.*, *10*, 221–68.
Vezina, C., & Singh, K. (1975). Transformation of Organic Compounds by Fungal
    Spores. In *The Filamentous Fungi*, vol. 1, ed. J. E. Smith & D. R. Berry,
    pp. 158–92. Edward Arnold, London.

Books and reviews detailing the use of immobilized microbial cells

Chibata, I., Tosa, T., & Sato, T. (1983). Immobilized Cells in the Formation of
    Fine Chemicals. *Adv. Biotechnol. Process*, *10*, 203–22.
Chibata, I., Tosa. T., & Sato, T. (1986). Methods in Cell Immobilization. In
    Manual of Industrial Microbiology and Biotechnology, ed. A. L. Demain &
    N. A. Solomon, pp. 217–29. American Society For Microbiology,
    Washington D.C.
Chibata, I., & Wingard, L. B. (1983). Immobilized Microbial Cells. *Adv. Biochem.
    Bioeng.*, *4*, 1–349.
Mattiason, B. (1983). *Immobilized Cells and Organelles*. Chemical Rubber Co.,
    Cleveland.

Books and reviews detailing methods of mutant strain selection

Demain, A. L. (1980). Increasing Enzyme Production by Genetic and Environmental
    Manipulations. *Meth. Enzymol.*, *22*, 119–36.
Elander, R. P., & Chang, L. T. (1979). Microbial Culture Selection. In *Microbial
    Technology*, 2nd ed., vol. 2, ed. H. J. Peppler & D. Perlman, pp. 243–302.
    Academic Press, New York.
Hopwood, D. A., & Merrick, M. J. (1977). Genetics of Antibiotic Production.
    Bacteriol. Rev., *41*, 595–634.

References

Abril, O., Ryerson, C. C., Walsh, C., & Whitesides, G. M. (1989). Enzymatic
    Baeyer-Villiger type oxidations of ketones catalysed by cyclohexanone
    oxygenase. *Bioorg. Chem.*, *17*, 41–52.
Alphand, V., Archelas, A., & Furstoss, R. (1990). Microbiological transformations.
    13. A direct synthesis of both *S* and *R* enantiomers of 5-hexadecanolide via an
    enantioselective microbiological Baeyer-Villiger reaction. *J. Org. Chem. 55*,
    347–50.
Anderson, M. S., Hall, R. A., & Griffin, M. (1980). Microbial metabolism of alicyclic
    hydrocarbons; cyclohexane catabolism by a pure strain of *Pseudomonas* sp. *J.
    Gen. Microbiol.*, *120*, 89–94.
Baum, R. H., & Gunsalus, I. C. (1962). Mono and bicyclic terpenoid oxidation and
    assimilation by a soil diphtheroid. *Bact. Proc.*, *108* (abstract, P31).
Chapman, P. J., Kuo, J.-F., & Gunsalus, I. C. (1963). Camphor oxidation:
    2,6-diketocamphane pathway in a diphtheroid. *Fed. Proc.*, *22*, 296.
Conrad, H. E., Dubus, R., & Gunsalus, I. C. (1961). Enzymatic lactonization of
    terpenes. *Fed. Proc.*, *20*, 48.

Hasegawa, Y., Hamano, K., Obata, H., & Tokuyama, T. (1982). Microbial degradation of cycloheptanone. *Agric. Biol. Chem.*, *46*, 1139–43.

Hayashi, T., Kakimoto, T., Ueda, H., & Tatsumi, C. (1970). Microbiological conversion of terpenes. Part VII. Conversion of borneol. *J. Agric. Chem. Soc., Japan*, *44*, 401–4.

Kay, J. W. D., Conrad, H. E., & Gunsalus, I. C. (1962). Camphor degradation: hydroxy intermediate formed by a soil diphtheroid. *Bact. Proc.*, *108* (abstract P30).

Khanchandani, K. S., & Bhattacharaya, P. K. (1974). Microbiological transformations of terpenes: Part XXI. Growth and adaptation studies on the degradation of camphene by the *Pseudomonas* camphene strain. *Ind. J. Biochem.*, *11*, 110–15.

Lee, S. S., & Sih, C. J. (1967). Mechanisms of steroid oxidation by microorganisms. XII. Metabolism of hexahydroindanpropionic acid derivatives. *Biochem.*, *6*, 1395–403.

Nakajima, O., Iriye, R., & Hayashi, T. (1978). Conversion of (−)-menthone by *Pseudomonas putida* strain YK-2.2. Metabolic intermediate and stereochemical structure of the metabolic products. *J. Agric. Chem. Soc., Japan*, *52*, 167–74.

Ouazzani-Chahdi, J., Buisson, D., & Azerad, R. (1987). Preparation of both enantiomers of a chiral lactone through combined microbiological reduction and oxidation. *Tet. Lett.*, *28*, 1109–12.

Shaw, R. (1966). Microbiological oxidation of cyclic ketones. *Nature*, *209*, 1369.

Shulka, O. P., Bartholomus, R. C., & Gunsalus, I. C. (1987). Microbial transformation of menthol and menthane-3,4-diol. *Can. J. Microbiol.*, *33*, 489–97.

Stirling, L. A., Watkinson, R. J., & Higgins, I. J. (1977). Microbial metabolism of alicyclic hydrocarbons; isolation and properties of a cyclohexane-degrading bacterium. *J. Gen. Microbiol.*, *99*, 119–25.

Tanaka, H., Shikata, K., Obata, H., Tokuyama, T., & Ueno, T. (1977). Isolation and cultural conditions of cyclohexanone-utilising bacterium. *Hakko-Kogaku Kaishi*, *55*, 57–61.

Trower, M. K., Buckland, R. M., Higgins, R., & Griffin, M. (1985). Isolation and characterization of a cyclohexane-metabolising *Xanthobacter* sp. *Appl. Environ. Microbiol.*, *49*, 1282–9.

Williams, D. R., Trudgill, P. W., & Taylor, D. G. (1989). Metabolism of 1,8-cineole by a *Rhodococcus* species; ring cleavage reactions. *J. Gen. Microbiol.*, *135*, 1957–67.

3

Useful intermediates and end-products obtained from whole-cell/enzyme-catalysed hydrolysis and esterification reactions

3.1 Introduction

Enzyme-catalysed hydrolysis and esterification reactions are the most commonly exploited biotransformations. There are two main reasons for this state of affairs. Firstly, the reactions are very easy to perform, and no special apparatus is required. Secondly, there is a wide range of hydrolase enzymes available from commercial suppliers. Furthermore, the stereo-selectivities and chemoselectivities of the enzymes are well known, and the likely stereochemical outcomes of such enzyme-catalysed reactions on previously unused substrates are now becoming predictable.

There are many types of enzyme-catalysed hydrolysis reactions. In this chapter, the hydrolysis of esters and amides will be surveyed quite extensively, while the hydrolysis of nitriles and epoxides will be mentioned briefly at the end. The use of hydrolase enzymes in organic solvents will be discussed also, in connection with the preparation of esters and amides.

In a number of instances, whole-cell preparations have been preferred as catalysts for selected hydrolysis reactions; in most of these cases the micro-organisms are easy to grow and simple to handle and are utilized because of the low cost involved. These cases will be integrated into the discussion as and when appropriate.

3.2 Hydrolysis of esters

The enzyme-catalysed hydrolysis of simple esters (Scheme 3.1) takes place at temperatures around 35°C and at pH values around 7 (i.e. under very mild conditions). Such reactions have been studied in great detail, and the ways in which the enzymes catalyse the reactions are being investigated. It is believed that some or all of these enzymes (called lipases or esterases) possess a serine residue as part of a "catalytic triad" of amino acids within

80

$$R^1CO_2R^2 + H_2O \xrightarrow{\text{enzyme}} R^1CO_2H + R^2OH$$

Scheme 3.1.

the active site of the enzyme. The serine residue is acylated by the ester, and the alcohol moiety is released into the solution. The acylated enzyme is then hydrolyzed by water to allow the carboxylic acid to move away from the active site, and the enzyme (being a true catalyst) reverts to its original state (Figure 3.1).

The enzyme is made up of (L)-amino acid residues, and hence it is a chiral catalyst. If the ester is chiral (i.e. it possesses one or more chiral centres in the alcohol moiety and/or the acid moiety), then there are two tetrahedral intermediates (A) that can be formed. These two intermediates have a diastereoisomeric relationship (Figure 3.2) and are different in energy (Figure 3.3). The two tetrahedral intermediates will be formed at different rates, the extent of the difference being a reflection of the energy gap ΔE; the overall biotransformation will be enantioselective to a greater or lesser extent.[1]

For example, the ester (1) is hydrolyzed by pig liver esterase (ple) to afford the acid (2) in optically active form and the recovered, optically active ester (Scheme 3.2). Of course, in order to obtain optically active products, the hydrolysis is not run to completion. Stopping the reaction at 40–50% conversion leads to isolation of the acid in a state of high optical purity. Contrary-wise, stopping the biotransformation at 50–60% conversion gives optically pure ester.

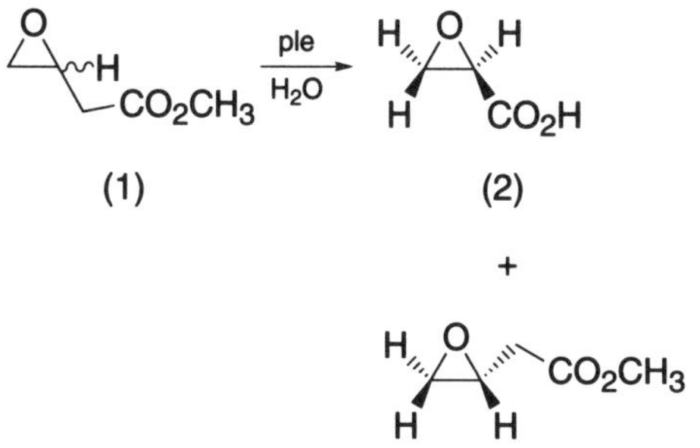

Scheme 3.2.

---

[1] It has been mooted that the stereoselectivity displayed by pig liver esterase is due, at least in part, to a difference in the rates of hydrolysis of the diastereoisomeric acyl enzyme complexes. In that case, the energy diagram would be modified, but the principle would remain the same.

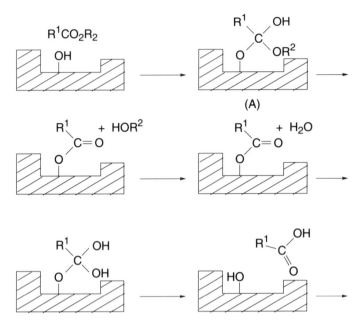

Figure 3.1. Schematic diagram showing the enzyme-catalysed hydrolysis of an ester.

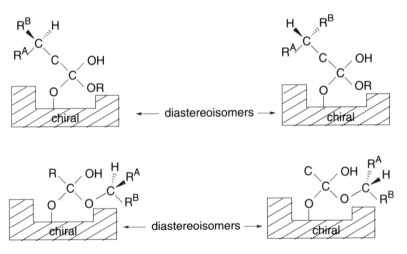

Figure 3.2. Schematic diagram showing the diastereoisomeric transition states formed on hydrolysis of a chiral ester using an enzyme.

The simple racemic ester ($\pm$)-3-acetoxyoct-1-yne (3) gave the (S)-alcohol (S)-(4) and the (R)-alcohol (R)-(4) in the ratio of 9:1 at ca. 50% conversion using *Mucor miehei* lipase (mml) as the catalyst (Scheme 3.3).

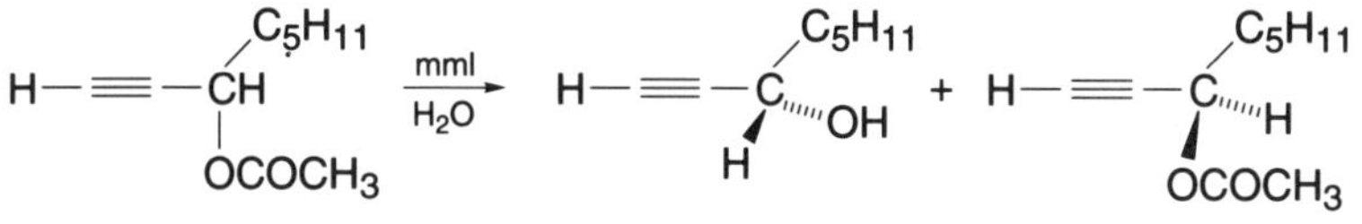

Scheme 3.3.

The "enantiomeric excess" (e.e.) of the (S)-alcohol is 80% [i.e. $(9-1)/(9+1) \times 100$]. The optically active alcohol (4) is a useful building block for the construction of various natural products, including prostaglandins and coriolic acid (Figure 3.4).

There are numerous examples of enantioselective hydrolyses of the types described in Schemes 3.2 and 3.3, catalysed by lipases and esterases. The selective hydrolysis of amino acid derivatives has been an important part of this field of study. For example, the hydrolase enzyme α-chymotrypsin catalyses the enantioselective hydrolysis of N-acetyl-DL-phenylalanine methyl ester (5) to give optically pure (L)-acid (6) in 40% yield (Scheme 3.4). The lipase from the fungus *Candida cylindracea* (ccl) has been shown to hydrolyse octyl 2-chloropropionate (7) with high stereoselectivity on a large scale, giving the (R)-acid (8) in 46% yield (96% e.e.) and the (S)-ester (45% yield) (Scheme 3.5).

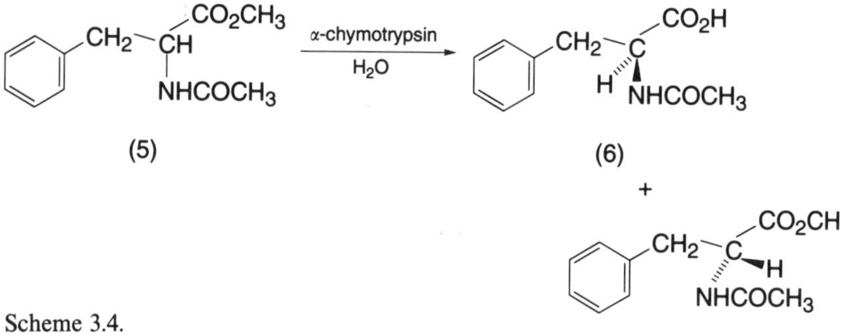

Scheme 3.4.

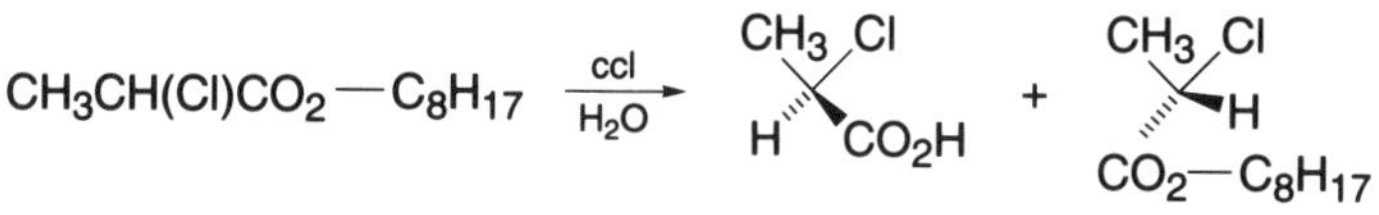

Scheme 3.5.

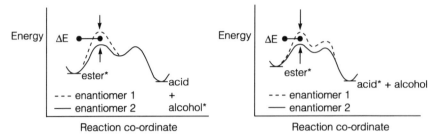

Figure 3.3. Energy diagrams relating to the hydrolase-catalysed fragmentation of chiral esters.

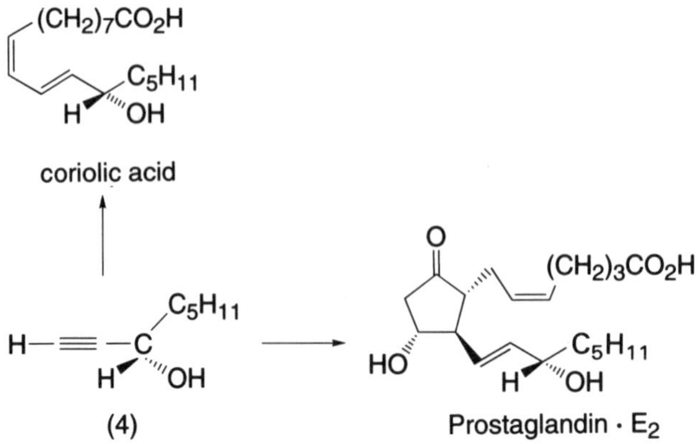

coriolic acid

(4)   Prostaglandin · E$_2$

Figure 3.4. Some of the uses of the alcohol (4) as an intermediate in organic synthesis.

The predictability of the outcomes of such hydrolyses (i.e. the stereo-chemistry of the more rapidly hydrolysed enantiomer) has been considerably advanced in recent years by the availability of active-site models for some of the more commonly used enzymes (e.g. pig liver esterase, porcine pancreatic lipase and *Candida cylindracea* lipase). These models have been constructed by analysing the stereoselectivity of the hydrolysis for a wide range of substrates. Recently, X-ray crystallographic data have been reported for two lipases, and such information obviously will lead to more accurate information on the active sites for these enzymes.

The formation of optically active synthons and the ready availability of such materials for synthesis are important because optically pure end-products often will be a prerequisite for incorporation into ethical drug formulations and important agricultural aids in the future. While one of the undoubted bonuses of using enzymes as asymmetric catalysts is the formation of optically pure chiral compounds, it should not be forgotten that the mild reaction conditions which can be employed mean that enzymes can be used to hydrolyze sensitive compounds, such as a prostaglandin derivative (9) and cephalosporins (10) (Scheme 3.6), which would decompose extensively under conventional conditions for ester hydrolysis (i.e. strong acid or alkali).

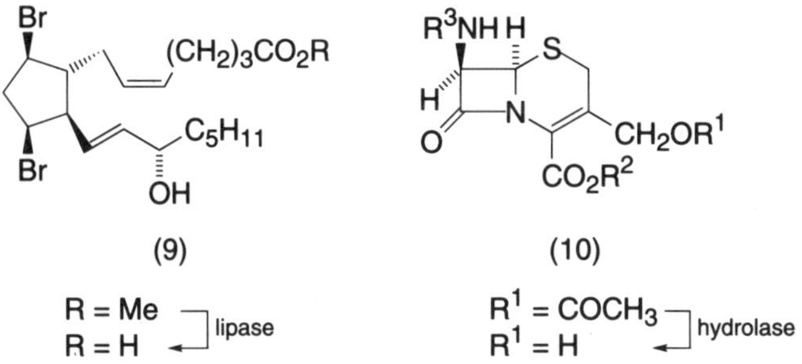

$R = Me$ ⌐ lipase
$R = H$ ⌐

$R^1 = COCH_3$ ⌐ hydrolase
$R^1 = H$ ⌐

Scheme 3.6.

In kinetic resolutions (Scheme 3.2–3.5) it is often the case that one of the products is required, while the other is not and must be discarded or recycled (e.g. racemised). Such operations can be wasteful or expensive. On the other hand, the biotransformation of *meso*-compounds or prochiral compounds allows for the possibility of preparing an optically pure compound in quantitative yield. In Scheme 3.7, two examples of the use of *meso*-compounds are described. The diester (11) is made up of a complex dicarboxylic acid unit derivatised as the dimethyl ester. Pig liver esterase catalyses the hydrolysis of one of the ester groups to give the acid (12) (95% e.e.) in 96% yield. This compound is an excellent precursor of the natural product neplanocin. Note that the acid (12) is not a substrate for ple, and thus the reaction stops at the "half-way" stage. The compound (13), like (11), possesses a plane of symmetry. Hydrolysis catalysed by porcine pancreatic lipase (ppl) affords the alcohol (14) (>98% e.e.) in quantitative yield. The latter compound has been used to make fluorocarbocyclic adenosine (C^F-adenosine), a stable analogue of the naturally occurring nucleoside adenosine.

The two examples shown in Scheme 3.7 fall into the appropriate categories following the guideline that esters of type (15) are preferentially hydrolysed by esterases, while compounds of type (16) are more susceptible to lipase-catalysed hydrolysis (Scheme 3.8).

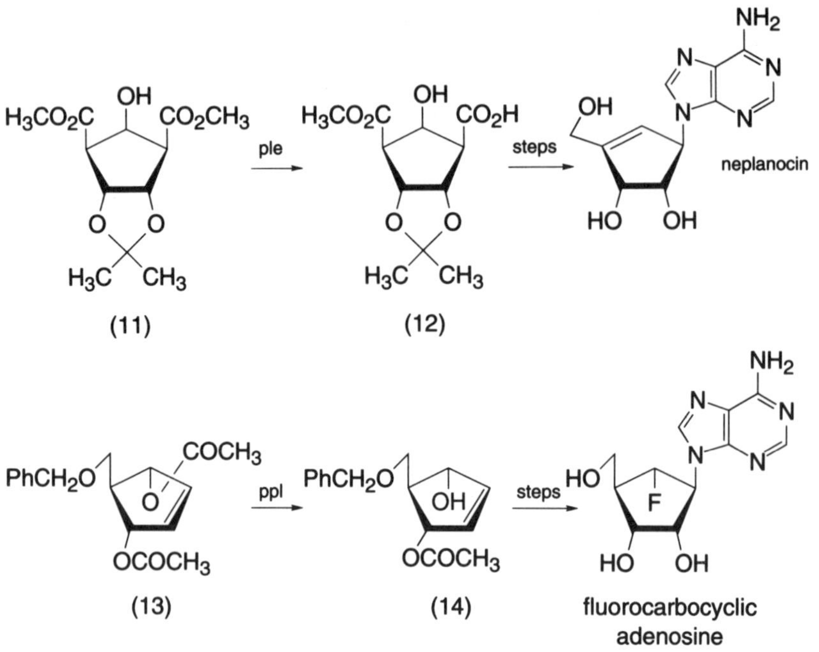

Scheme 3.7.

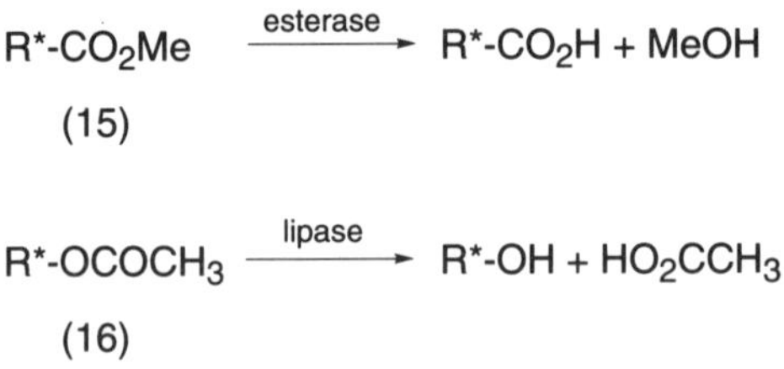

$$\text{R*-CO}_2\text{Me} \xrightarrow{\text{esterase}} \text{R*-CO}_2\text{H} + \text{MeOH}$$

(15)

$$\text{R*-OCOCH}_3 \xrightarrow{\text{lipase}} \text{R*-OH} + \text{HO}_2\text{CCH}_3$$

(16)

Scheme 3.8.

As adumbrated earlier, the hydrolysis of *meso*-compounds or prochiral compounds can provide optically active intermediates, useful for the synthesis of pharmaceuticals and other high-value materials. A further example is provided by the prochiral diester (17), which is hydrolysed using ple as the catalyst to give the chiral acid (18) (93% e.e., 93% yield). The protected amino acid (18) was converted through a series of conventional chemical steps into the anti-bacterial agent thienamycin. Similarly, the diester (19) provides the hydroxyester (20) on hydrolysis utilizing ppl as the catalyst. Over-reaction can be a problem in this case, and a sample of the alcohol (20) with good optical purity (93% e.e.) was obtained only if the reaction was curtailed at ca. 30% conversion (Scheme 3.10). A good deal of thought has gone into providing computer-based models for such biotransformations (i.e. when two or more reactions with different kinetic parameters are involved).

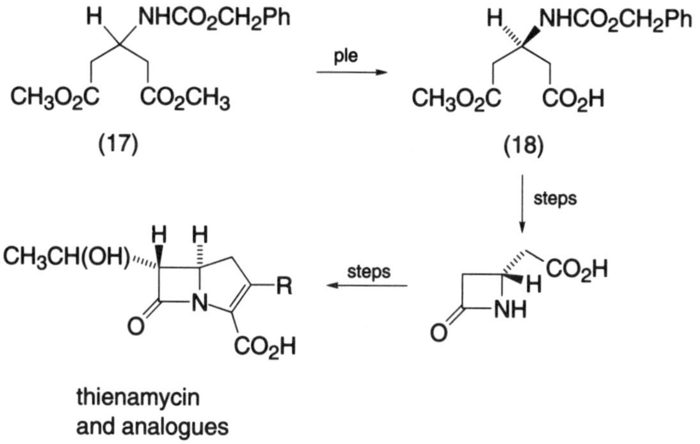

thienamycin
and analogues

Scheme 3.9.

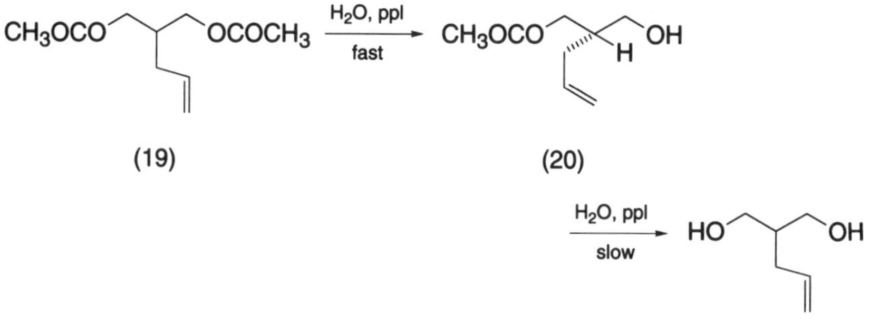

Scheme 3.10.

3.3 Hydrolysis of amides

The hydrolysis of amide bonds in protein and simpler entities by enzymes has been known for many years and explored extensively. The degradation of protein by a proteinase such as subtilisin is the basis of action of "biological" washing powders. Obviously the commercial importance of subtilisin has focussed much attention on the enzyme, and it has been modified by genetic engineering to provide a catalyst that will operate at various temperatures (viz. in the "cold wash" and in the "hot wash").

α-Chymotrypsin, one of the first enzymes to be investigated for use in preparative chemistry, also catalyses the hydrolysis of amide bonds of proteins. The cleavage is rapid when the carbonyl group of the amide bond is part of an aromatic amino acid residue. It should be emphasized that the catalytic activity of α-chymotrypsin is not restricted solely to amide hydrolysis; a wide range of esters can also be hydrolysed using this inexpensive enzyme (e.g. Scheme 3.4).

The mild reaction conditions that can be employed in the cleavage of such amide linkages means, once again, that sensitive bonds can be preserved. This is relevant to the commercially important process which converts fermentated penicillins (e.g. penicillin G) into 6-aminopenicillanic acid (6-APA), a precursor of the semi-synthetic penicillin antibiotics. The enzyme used for this transformation is the acylase from the bacterium *Escherichia coli* (Scheme 3.11) (see also Chapter 6, Section 6.5).

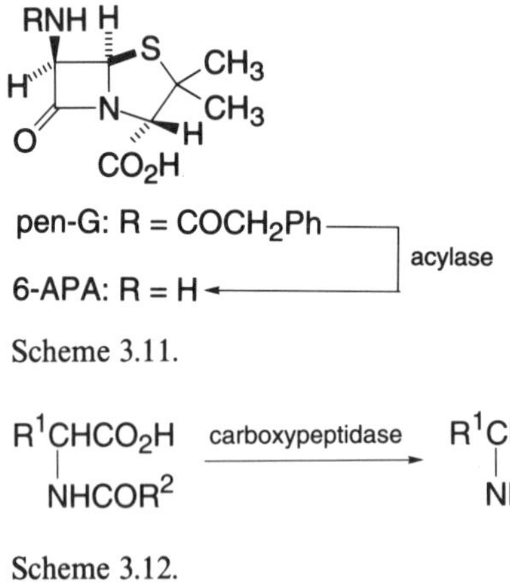

pen-G: R = COCH$_2$Ph ─┐
                        ├ acylase
6-APA: R = H ◄──────────┘

Scheme 3.11.

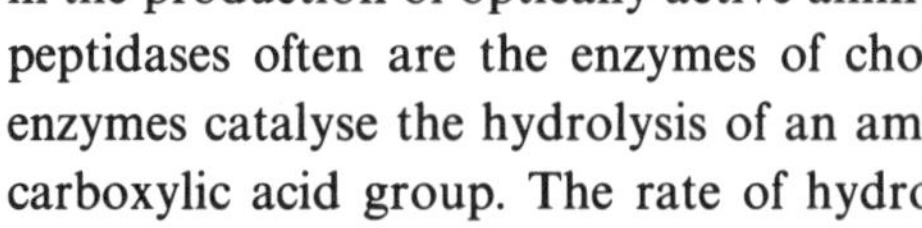

Scheme 3.12.

Enantioselective cleavage of non-peptide amide bonds is also important in the production of optically active amino acids (Scheme 3.12). Carboxypeptidases often are the enzymes of choice in this area of work; these enzymes catalyse the hydrolysis of an amide function which is close to a carboxylic acid group. The rate of hydrolysis is usually increased if R^1 (Scheme 3.12) is an aromatic unit or a large aliphatic moiety. For example, *threo-β*-phenylserine $R^1 = $ PhCH(OH) has been resolved by incubation of the racemic N-trifluoroacetate with carboxypeptidase-A, with the optically pure (L)-enantiomer being obtained in a good yield.

A closely related method for the preparation of optically pure amino acids involves the formation of hydantoins and the use of hydantoinases (Scheme 3.13). One advantage of this method is that when R is an aromatic group, in situ racemisation of the substrate can take place, leading to a high yield ($\gg 50\%$) of optically pure amino acid (for further details, see Section 6.4.2). It is noteworthy that optically active amino acids can be prepared utilizing esterases, amidases (acylases) or hydantoinases; six or seven amino acids are made commercially using one or another of these biotransformation processes.

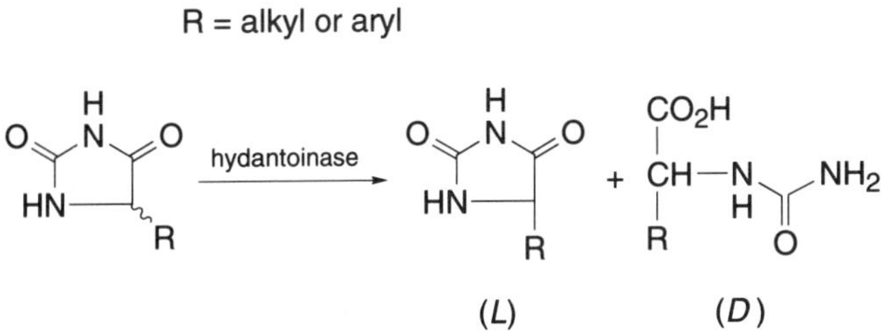

R = alkyl or aryl

Scheme 3.13.

The amide unit to be cleaved by an enzyme can also be part of a cyclic system; thus lactams can also be hydrolysed selectively using enzymes or micro-organisms. For example, the γ-lactam (22) (Scheme 3.14) is hydrolysed enantiospecifically by an amidase in *Rhodococcus equi* to give the amino

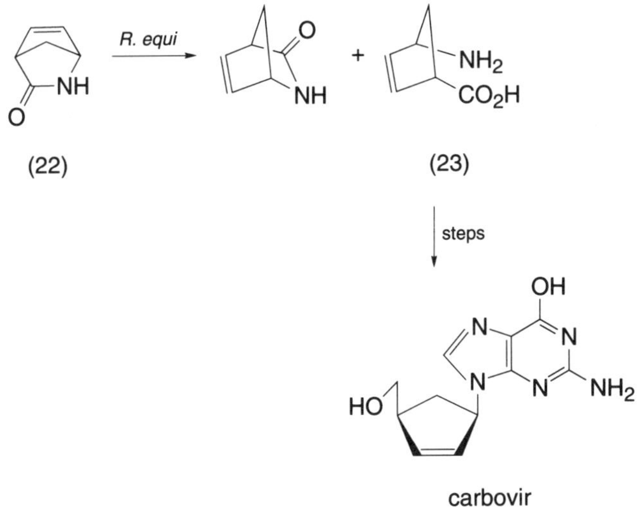

carbovir

Scheme 3.14.

acid (23) and the recovered lactam, both, essentially, optically pure (Scheme 3.14). The non-natural amino acid (23) is a useful building block, allowing the cyclopentene derivative carbovir to be made in an optically active form. Carbovir has been under intense scrutiny as a potential medicine, due to its pronounced activity against human immunodeficiency virus (HIV). *Rhodococcus equi* can also be used to hydrolyse the β-lactam (24) in an enantiospecific fashion (Scheme 3.15). The β-amino acid (25) is obtained together with recovered, optically pure lactam. The latter compound is used to prepare the natural product cispentacin. This cyclopentane derivative displays quite remarkable activity against *Candida* infections in vivo, and it, or a derivative, may provide the basis for a new clinical treatment for fungal infections such as oral or vaginal thrush.

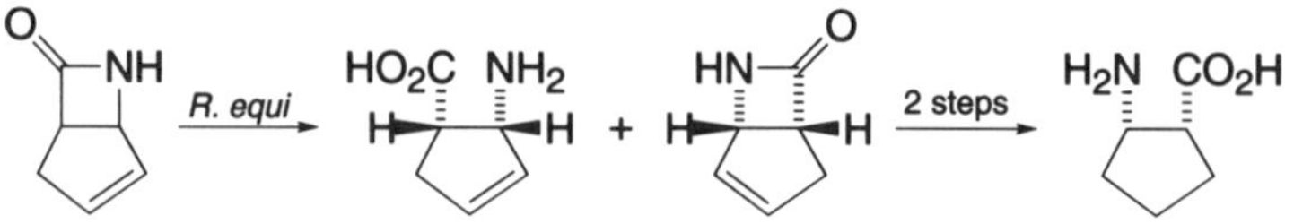

| (24) | | | (25) |
| | | | cispentacin |

Scheme 3.15.

3.4 Enzyme-catalysed formation of esters and amides

It was a surprise to many to find that some enzymes are *not* rapidly denatured by organic solvents, and, indeed, in the mid-1980s a team from the Massachusetts Institute of Technology popularized the use of lipases in organic solvents. The reaction medium often contains just enough water to form an aqueous coat around the enzyme. Solvents such as dichloromethane, hexane, and toluene can be employed in this work. The enzyme can then be made to "work backwards", that is, to couple an acid and an alcohol to give an ester. The reverse reaction can be enantioselective: Thus, on using *Candida cylindracea* lipase (ccl) in hexane containing a trace quantity of water, it was found that a wide variety of α-halogenated acids (26) reacted with alcohols such as *n*-butanol to yield both the optically active ester (27) and the unreacted enantiomer of the acid (Scheme 3.16).

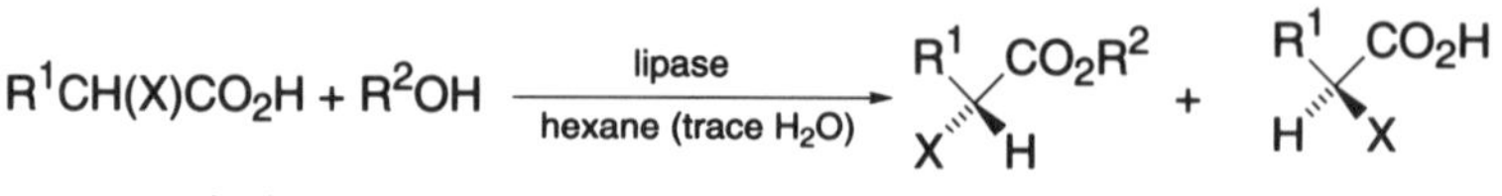

$$R^1CH(X)CO_2H + R^2OH \xrightarrow[\text{hexane (trace H}_2\text{O})]{\text{lipase}} \underset{(27)}{\overset{R^1 \quad CO_2R^2}{\underset{X \quad H}{\bigtimes}}} + \underset{H \quad X}{\overset{R^1 \quad CO_2H}{\bigtimes}}$$

(26)

Scheme 3.16.

The lipase is lyophilized (most of the water is removed from around the protein) before use in such a system, to avoid a pronounced back-reaction (hydrolysis). If the enzyme shows different catalytic properties at different pH values, then these differences will be reflected in the reactions taking place in the organic solvent. The enzyme "memorizes" the pH of the solution from which it was lyophilized!

The enzyme porcine pancreatic lipase (ppl) displays low catalytic activity in such esterification reactions. However, this enzyme was found to be active and extremely stereoselective for *trans*-esterification reactions in anhydrous organic systems. A practical use for such a *trans*-esterificaton process involves the resolution of the pheromone sulcatol (28) (Scheme 3.17).

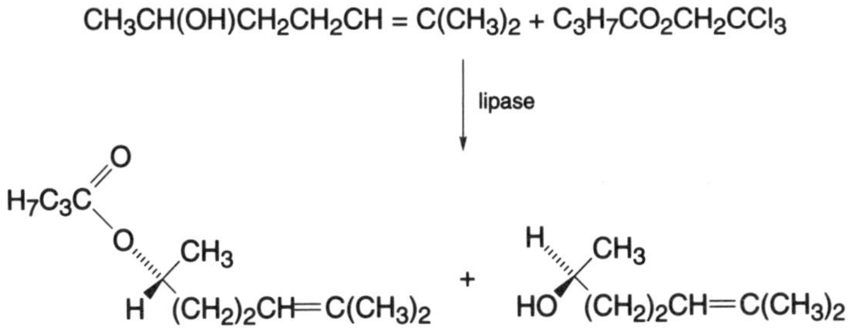

$$CH_3CH(OH)CH_2CH_2CH = C(CH_3)_2 + C_3H_7CO_2CH_2CCl_3$$

Scheme 3.17.

A related inter-esterification reaction has been employed to demonstrate the concept of "double enantioselection" (Scheme 3.18). Thus the bicyclic

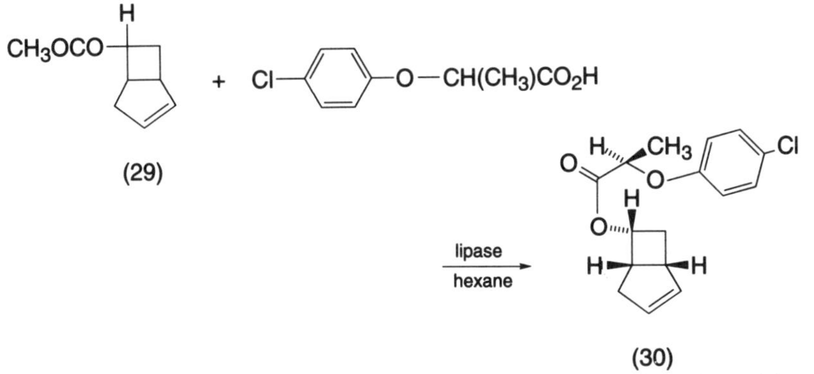

Scheme 3.18.

ester (29) and *p*-chlorophenoxypropanoic acid in hexane containing a very small quantity of water were coupled using *Candida cylindracea* lipase as the catalyst to give the ester (30) as the major diastereoisomer formed (91%). The three other diastereoisomers were each produced to the extent of about 3%. It is interesting to note that such an inter-esterification reaction is considerably more selective than the corresponding direct esterification reaction because, using the strategy outlined in Scheme 3.18, the bicyclic unit has to visit the enzyme's active site on two occasions; the first visit serves to remove the acetate group, and the second visit is necessary to react the so-formed alcohol with the bulky acyl unit (Scheme 3.19).

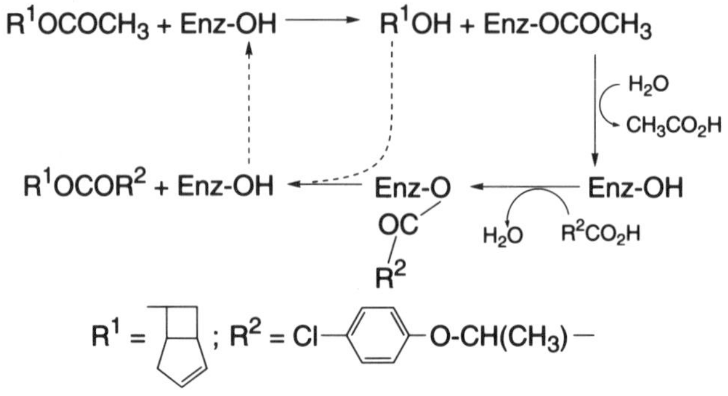

Scheme 3.19.

A modification of the *trans*-esterification process has found favour with synthetic chemists. The use of vinyl acetate (often as the solvent) for the acetylation of chiral secondary alcohols leads to the formation of vinyl alcohol, which rapidly tautomerizes to acetaldehyde. Acetaldehyde does not take part in a back-reaction (but can form a Schiff base derivative with the enzyme, which may affect the catalyst's activity). For example, the alcohol (31) is readily formed from cyclopentadiene and is enantiospecifically acylated under the influence of *Pseudomonas fluorescens* lipase (Scheme 3.20) to furnish the acetate (32) and the recovered, optically active alcohol. The unreacted alcohol (−)-(31) was chemically acetylated and converted into the biological active carbocyclic nucleoside aristeromycin. For another example involving this protocol, see Chapter 5 (p. 133).

Similarly, an amine and a carboxylic acid can be condensed together to form an amide using a hydrolase enzyme "working backwards". The method has been used widely to prepare small peptides. The preparation of the sub-sequence of dynorphin (33) was accomplished using a medley of enzymes, namely α-chymotrypsin (α-C), papain (P) and trypsin (T)

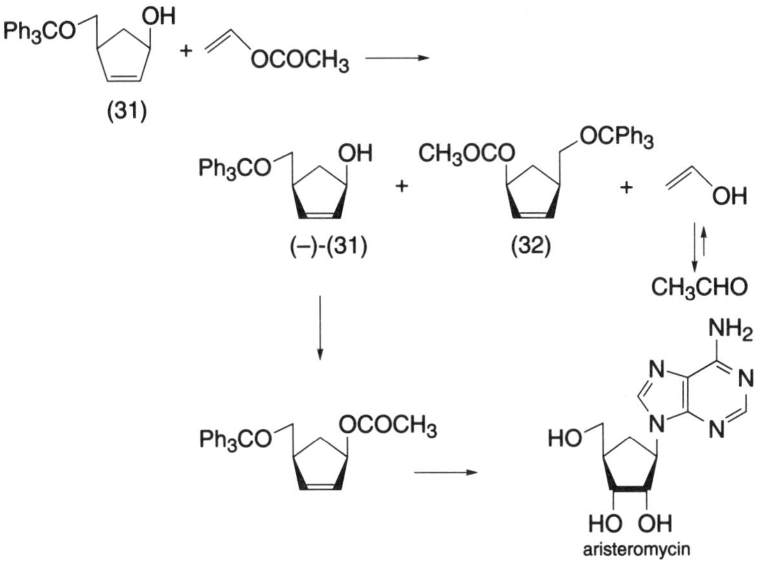

Scheme 3.20.

(Scheme 3.21). (Note that the protecting groups that had to be employed in Scheme 3.21 have been omitted for the sake of clarity.) The enzyme catalysed

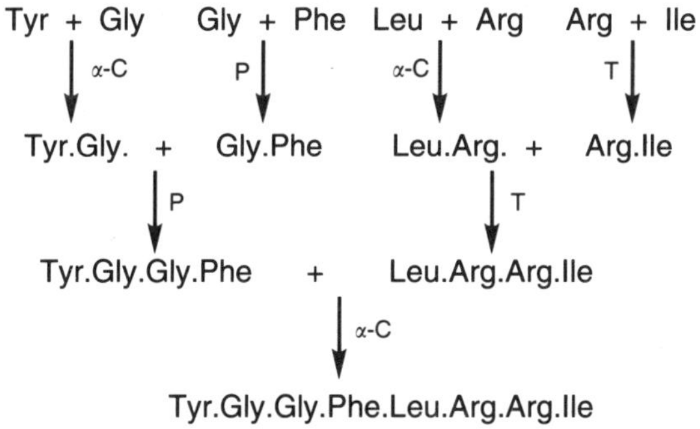

Scheme 3.21.

coupling of an amine and an acid has also been explored in connection with the preparation of the artificial sweetener Aspartame. Immobilized thermolysin was used as the catalyst to couple (L)-phenylalanine methyl ester with N-carbobenzoxy-(L)-aspartic acid to give the desired product (34). The side-chain carboxyl group of the aspartic acid derivative does not have to be protected because thermolysin is specific in its catalytic activity.

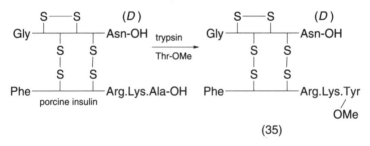

(34)

It has also been demonstrated that lipases can be used to make amide bonds. Thus both porcine pancreatic lipase and *Candida cylindracea* lipase, in an organic solvent, proved to be effective catalysts for the synthesis of a wide range of dipeptides.

A commercially viable preparation of the ester of human insulin (35) has been developed using a trypsin-mediated exchange of a threonine residue for the terminal alanyl residue of porcine insulin (Scheme 3.22). This transformation was the first example of an enzymic semi-synthesis of a protein for use in medicine. (See Section 6.9.2 for further details on the production of Aspartame and insulin.)

Scheme 3.22.

3.5 Hydrolysis of nitriles and epoxides

The hydrolysis of aliphatic and aromatic nitriles into amides and/or carboxylic acids is an area of rapidly increasing interest. Already the Nitto Chemical Industry Co. Ltd. has developed a large-scale process for the conversion of acrylonitrile into acrylamide using *Rhodococcus* cells (see Section 6.7.2). Many other organisms, as well as the nitrilases, nitrile hydratases and amidases derived from them, can effect similar conversions. It is interesting to note that for dinitriles, only one of the nitrile groups is hydrolysed (Scheme 3.23) (the polar cyanocarboxylic acid is not a substrate for the enzyme). In addition, it is clear that a nitrile group is hydrolysed (to an amide or carboxylic acid moiety) in preference to an ester group (Scheme 3.23), a situation that is difficult to emulate using conventional, purely chemical hydrolysis conditions.

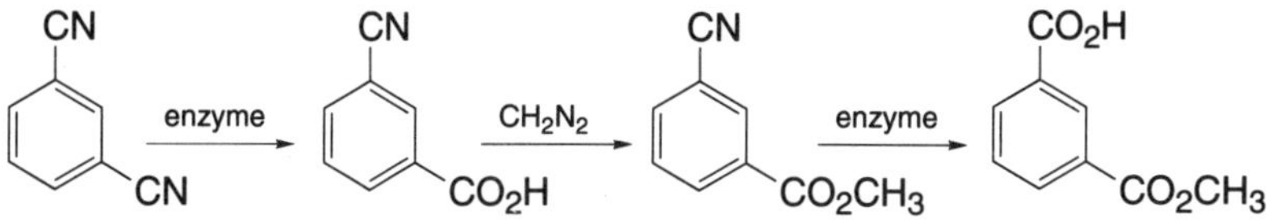

Scheme 3.23.

The first cases of enantioselective hydrolyses of chiral nitriles are emerging (Scheme 3.24), as are the selective hydrolyses of prochiral dinitriles, to give potentially useful, optically active ω-cyanocarboxylic acids (Scheme 3.24).

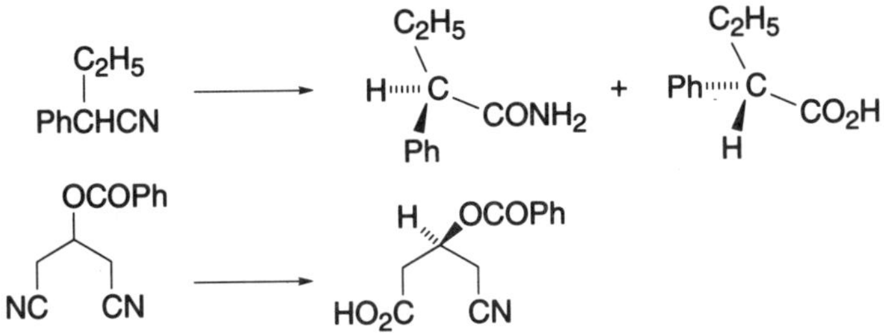

Scheme 3.24.

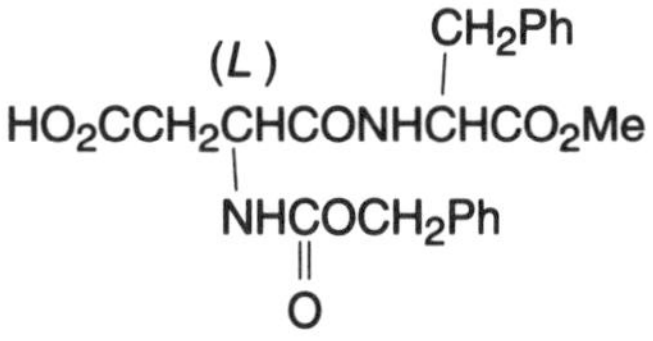

(34)

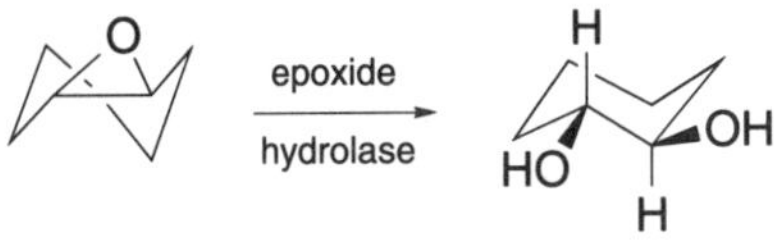

Scheme 3.25.

The hydrolysis of epoxides to give 1,2-diols is an area that is ripe for development. Some work has been published showing that epoxides such as cyclohexane epoxide (36) form optically active diols, in this case cyclohexane-(1R,2R)-diol (37). The research has concentrated on the use of enzymes present in liver microsomes, and while this elegant work has indicated what can be achieved, it is clear that rapid progress and the involvement of non-experts in this particular area must await the discovery of readily available epoxide hydrolase enzyme(s) from microbial sources.

3.6   Conclusions and overview

The hydrolyses of esters, amides, and nitriles involve very simple procedures that are easy to perform, even by the non-specialist. Similarly, the formation of esters using lipases is a process that can be considered to be absolutely straightforward. The reactions can be conducted under very mild conditions, and the regioselectivities and stereoselectivities that can be obtained are such that these procedures are becoming quite commonplace for the production of chemically and optically pure intermediates and end-products.

The optical purity of a compound, be it an intermediate or end-product, can be expressed as the enantiomeric excess. It is noteworthy that during a kinetic resolution $[S \Rightarrow S^* + P^*]$, the enantiomeric excess of the product (P^*) or recovered substrate (S^*) will depend on the extent of conversion (c). Charles Sih introduced the term "enantiomeric ratio", E, which is independent of the degree of conversion and is a very useful way of expressing the selectivity of an irreversible transformation. The value for E is calculated as follows:

$$E = \frac{\ln\left[(1-c)(1-\text{e.e.}_S)\right]}{\ln\left[(1-c)(1+\text{e.e.}_S)\right]}$$

where e.e.$_S$ is the enantiomeric excess of the substrate. E values 0–10 indicate that a transformation has low selectivity, E values 10–100 indicate moderate selectivity, and E values greater than 100 reflect good selectivity.

It should be emphasized that the examples cited in this chapter are typical of many hundreds of biotransformations that have appeared in the literature over the past 10 years. Texts featuring comprehensive reviews of the subject matter that has been outlined in this chapter are listed in the Bibliography.

Bibliography

Boland, W., Frossl, C., & Lorenz, M. (1991). Esterolytic and Lipolytic Enzymes in Organic Synthesis. *Synthesis*, 1049.

Faber, K. (1992). *Biotransformations in Organic Chemistry*. Springer-Verlag, Berlin.

Faber, K., & Riva, S. (1992). Enzyme-catalysed Irreversible Acyl Transfer. *Synthesis*, 895.

Klibanov, A. M. (1990). Asymmetric Transformations Catalysed by Enzymes in Organic Solvents. *Acc. Chem. Res., 23*, 114.

Rozell, J. D., & Wagner, F. (1992). *Biocatalytic Production of Amino-Acids and Derivatives*. Hanser Publishers, New York.

Tramper, J., Vermuë, M. H., Beeftink, H. H., & Von Stockar, U. (1992). *Biocatalysis in Non-conventional Media*. Elsevier Science, Amsterdam.

Santaniello, E., Ferraboschi, F., Grisenti, P., & Manzocchi, A. (1992). The Biocatalytic Approach to the Preparation of Enantiomerically Pure Chiral Building Blocks. *Chem. Rev., 92*, 1071.

4

Useful intermediates and end-products obtained from biocatalysed oxidation and reduction reactions

4.1 Introduction

As discussed in Chapter 1, there has been much interest, over many years, in oxidation and reduction reactions catalysed by micro-organisms. Processes for alcohol production and for the manufacture of vinegar and vitamin C have intrigued investigators for ages. Today there is great interest in discovering new whole-cell and isolated-enzyme-catalysed reactions of this type. As described in Chapter 3, the focus of the present-day work in this area is the production of compounds (often in optically active form) that might be useful as intermediates or end-products for the pharmaceutical, fragrance, flavour and agrichemical industries.

For many oxidation and reduction reactions it is often necessary to make a choice between using a whole-cell system or an isolated enzyme (with the requisite co-factors) for the biotransformation. In this area, factors that influence the choice often are much more complicated than those for hydrolysis reactions. The pros and cons of employing whole cells or partially purified oxidoreductase enzymes have been discussed in Chapter 2, and the salient features are summarized in Table 2.3. While the advantages and disadvantages will not be discussed in detail again, many of the points made earlier will recur in this discussion.

4.2 Reduction of ketones using whole cells

For example, consider one of the simplest reduction processes, namely the reduction of a ketone to a secondary alcohol. Of course, this can be accomplished chemically using sodium borohydride. The same transformation can be achieved using baker's yeast, and one advantage of using the biocatalyst often can be seen immediately (i.e. optically active forms of chiral alcohols can be obtained). The reason is simple: The bio-reduction

takes place in a chiral cavity of an enzyme. For example, cyclohexyltri-
fluoromethyl ketone (1) is readily reduced to the alcohol (2) very efficiently.
Similarly, ethyl 3-oxobutanoate forms ethyl 3(*S*)-hydroxybutanoate with
very high selectivity using the same biocatalyst (Scheme 4.1). Baker's yeast

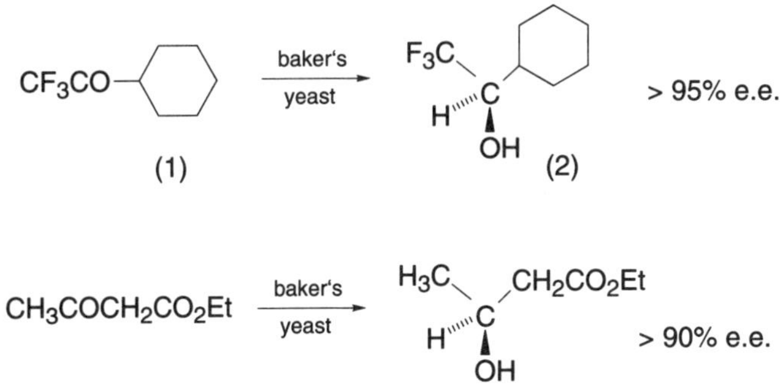

Scheme 4.1.

has also been used to reduce dicarbonyl compounds, such as cyclopentane-
1,3-dione derivatives. If the carbon atom between the two carbonyl groups
carries two different substituents, then the starting material (a prochiral
compound) often is converted in very high yield and excellent stereoselectivity to a compound possessing two chiral centres (Scheme 4.2). If

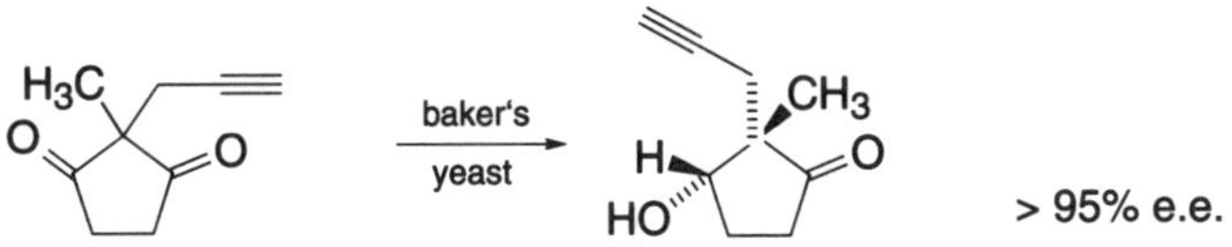

Scheme 4.2.

the starting ketone or ketoester has a pre-existing chiral centre(s), then
enantioselective reduction of the racemic starting material can give a
product rich in one diastereoisomer (Scheme 4.3); alternatively, both
enantiomers of the starting material can be reduced to give an equimolar
mixture of two diastereoisomers (Scheme 4.3).

The outcome of a yeast-catalysed reduction of a ketone can be predicted
with some confidence. For simple dialkyl or aralkyl ketones, the product
obtained is almost invariably the (*S*)-alcohol (Figure 4.1).

While the yeast reduction of β-ketoesters can appear, at first sight, to
be more unpredictable (Scheme 4.4), the results can be interpreted using
a simple model (Figure 4.2) and will depend on whether the (substituted)

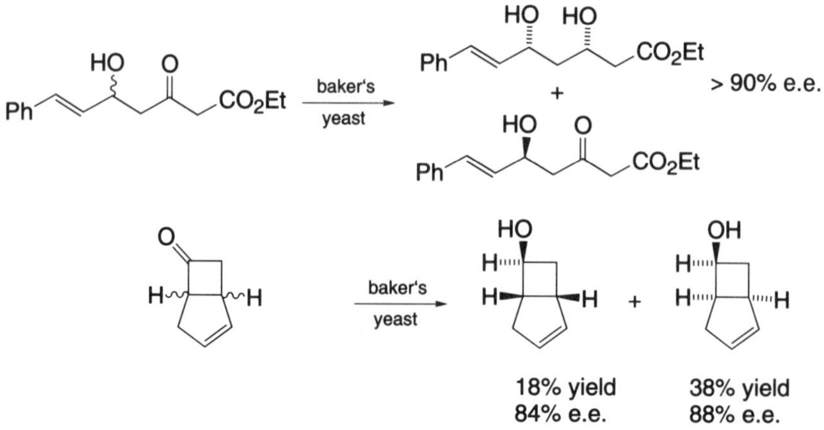

Scheme 4.3.

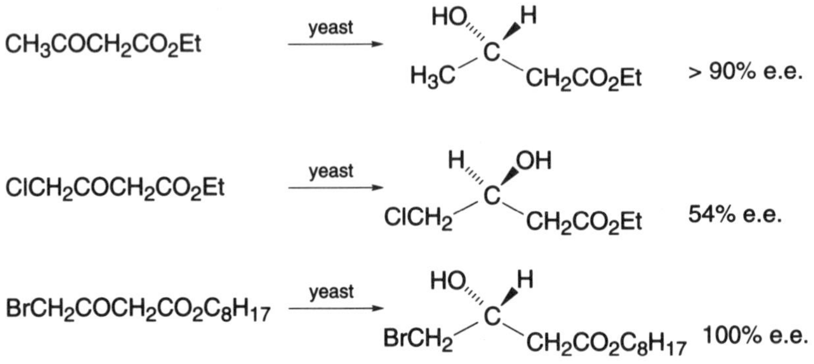

Scheme 4.4.

methyl group or the acetic acid residue is the more bulky moiety (CH_3, small; CH_2CO_2Et, medium; $ClCH_2$ or $BrCH_2$, large; $CH_2CO_2C_8H_{17}$, very large).

Yeast-catalysed reduction of 2-substituted β-ketoesters often gives rise to diastereomers, both possessing the (S)-configuration at the newly formed chiral centre. The ratio of the diastereomers often is not 50:50, since enolisation, in situ racemisation, and the preferential reduction of one of the enantiomers lead to a predominance of one of the stereoisomers (Scheme 4.5).

The ready availability of baker's yeast and the ease of operation of the whole-cell reduction are two of the attractive features of these processes. Many other organisms can be used to reduce ketones to secondary

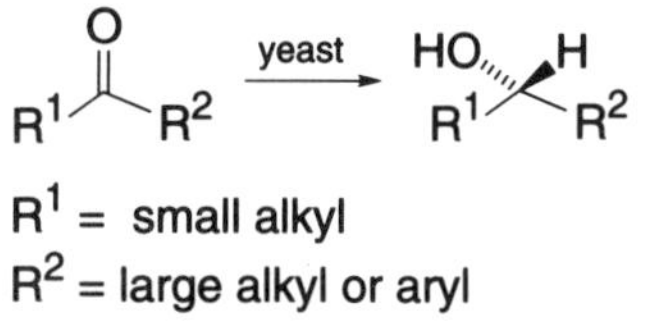

R^1 = small alkyl
R^2 = large alkyl or aryl

Figure 4.1. Reduction of unsymmetrical ketones with yeast.

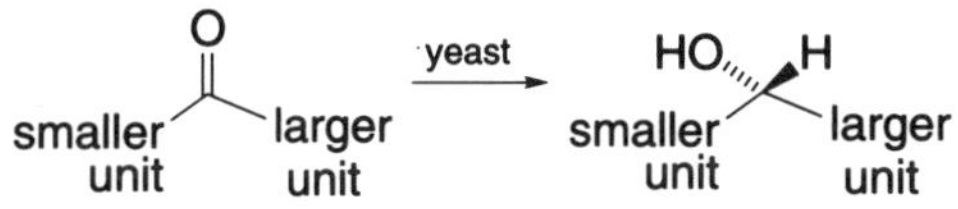

Figure 4.2. Yeast reduction of ketones with substituents of differing sizes.

alcohols. Some fungi (such as *Mortierella isabellina*) give stereochemical results broadly in line with those found for baker's yeast, while others (e.g. *Aspergillus niger*) give the opposite stereoisomer to that formed from the corresponding yeast reduction (Scheme 4.6).

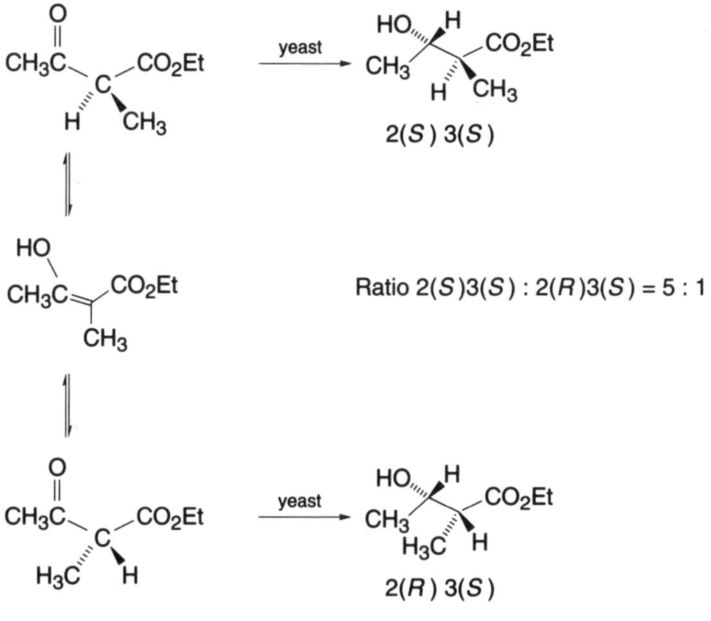

Ratio 2(*S*)3(*S*) : 2(*R*)3(*S*) = 5 : 1

Scheme 4.5.

It is interesting to note that large quantities of 3(*R*)-hydroxyesters of type (3) can be obtained from a variety of micro-organisms which store the

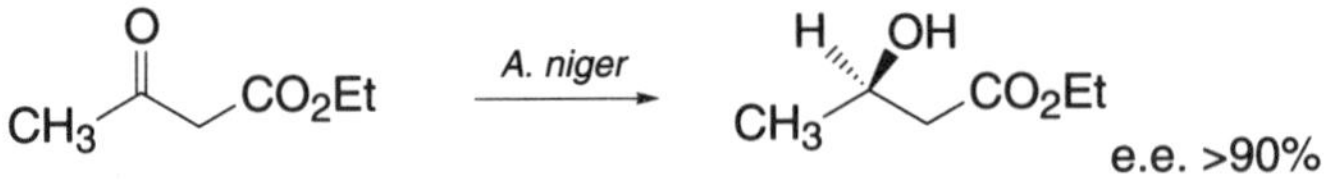

Scheme 4.6.

material as polyhydroxybutyrate (PHB). The polymer is converted into the monoester in vitro using sulphuric acid in the requisite alcohol.

4.3 Reduction of carbon–carbon double bonds

Yeast and, to a lesser extent, other micro-organisms have been used to accomplish the reduction of carbon–carbon double bonds. The susceptible double bonds are those conjugated to electron-withdrawing groups. Thus α-substituted α,β-unsaturated acids can be converted into chiral compounds with high optical activity. Note that the configuration about the alkene bond can determine the configuration of the newly formed chiral centre (Scheme 4.7). α,β-Unsaturated aldehydes are reduced in a similar fashion

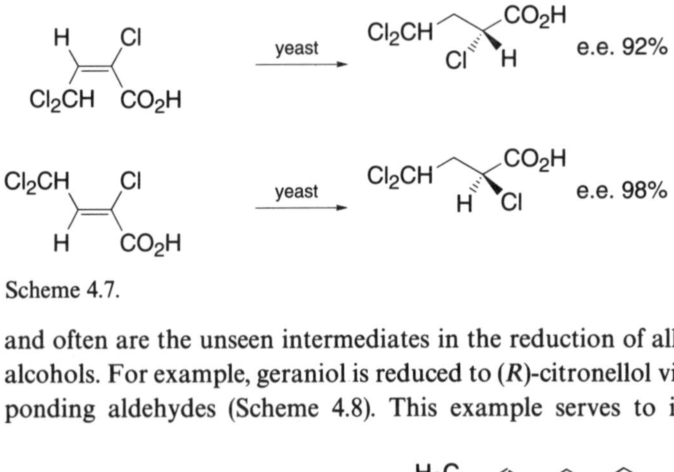

Scheme 4.7.

and often are the unseen intermediates in the reduction of allylic primary alcohols. For example, geraniol is reduced to (R)-citronellol via the corresponding aldehydes (Scheme 4.8). This example serves to illustrate the

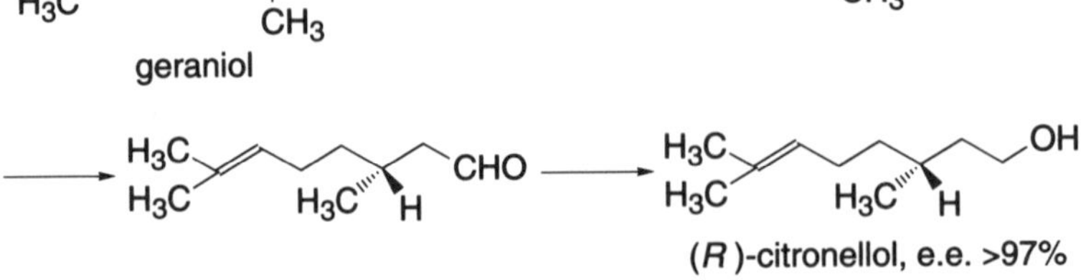

Scheme 4.8.

chemoselectivity of the reduction process: The remote alkene unit is not transformed in this process – similarly with dienoic acids or esters, only the α,β-bond is reduced.

The chiral cyclohexane-1,4-dione derivative (5), which is used in the synthesis of various naturally occurring carotenoids, is obtained by reduction of the corresponding ene-dione (Scheme 4.9). This biotransformation has been scaled up to the multi-kilogram level.

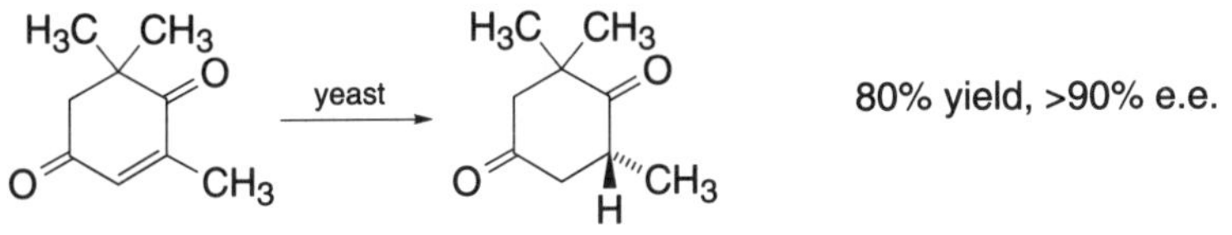

Scheme 4.9.

4.4 Reduction of ketones and α,β-unsaturated carbonyl compounds using enzymes

All the biotransformations discussed in this chapter thus far have involved whole cells (usually baker's yeast). It is quite straightforward to isolate the dehydrogenase enzymes[1] that are involved in the conversion of a ketone to a secondary alcohol. These enzymes can be obtained from micro-organisms (e.g. yeast) or animal sources (e.g. horse liver). During the purification process, the requisite co-factor, nicotinamide-adenine dinucleotide (phosphate) (NAD(P)), is disengaged from the protein. To reconstitute the catalytic power of the enzyme, the co-factor must be replaced. It is impossible to use a stoichiometric amount of co-factor, for several reasons, one principal factor being the high cost of NAD(P). Hence a less-than-stoichiometric amount of co-factor is used, and co-factor recycling is employed (Figure 4.3).

The question of co-factor recycling has been addressed by a number of scientists, and several convenient solutions to the problem have been found. One of these solutions is shown in Figure 4.4, which features the working "head group" of the co-factor; the figure demonstrates the use of a second enzyme, formate dehydrogenase, which takes the co-factor back to the reduced form, with the evolution of carbon dioxide.

A least aesthetic but perfectly workable method involves adding an excess of a low-molecular-weight (i.e. volatile) sacrificial alcohol such as

---

[1] The enzymes catalyse both the reduction of ketones to alcohols and the reverse process (i.e. the oxidation of secondary alcohols to ketones) under appropriate conditions. The name of the class of enzymes emphasizes the latter capability.

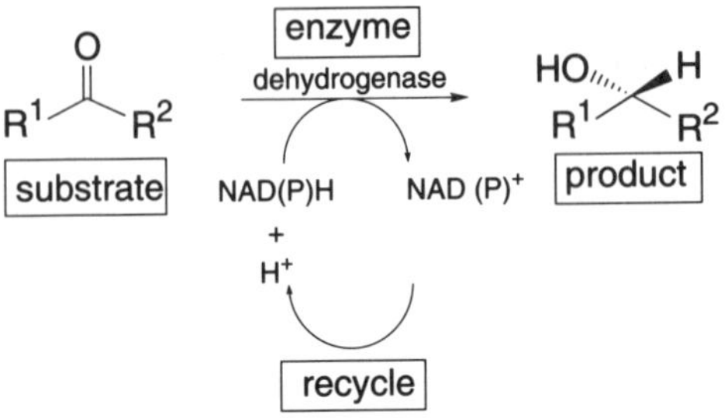

Figure 4.3. Recycling NAD(P)H in enzyme-catalysed reductions.

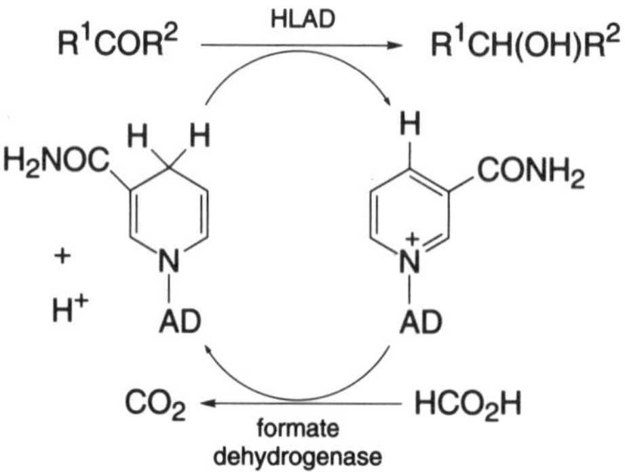

Figure 4.4. Molecular detail regarding role of NAD(P)H in dehydrogenase-catalysed reductions.

ethanol or isopropanol and allowing the dehydrogenase to work both reactions (Figure 4.5). The dehydrogenase illustrating the latter methodology (TBAD) was isolated from a micro-organism that thrives near volcanic hot springs; in such organisms, the enzymes, of necessity, are stable to temperatures of 80 °C or more.

The use of isolated enzymes and the requisite co-factor means that one of the major drawbacks of employing whole-cell systems, namely the occurrence of side-reactions (see Section 2.3.5), is circumvented. Thus yeast cells contain enzymes that catalyse hydrolysis, oxidation/reduction, and many other reactions. Dehydrogenase enzymes are catalysts for oxidation and reduction processes only. Figure 4.6 shows that yeast alcohol dehydrogenase (YAD), *Thermoanaerobium brockii* alcohol dehydrogenase (TBAD) and horse liver alcohol dehydrogenase (HLAD) complement each other in terms of the preferred sizes of the substrates: High-molecular-weight

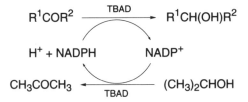

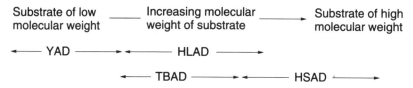

Figure 4.5. Recycling of NADPH using isopropanol.

| Substrate of low molecular weight | ——— | Increasing molecular weight of substrate | ———→ | Substrate of high molecular weight |

←——— YAD ————→ ←——— HLAD ———→

←——— TBAD ———→ ←—— HSAD ·———→

Figure 4.6. Substrate ranges for four dehydrogenases.

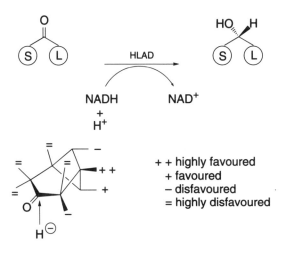

+ + highly favoured
+ favoured
– disfavoured
= highly disfavoured

Figure 4.7. Reduction of some cyclic ketones.

ketones are best handled by 3α, 20β-hydroxysteroid alcohol dehydrogenase (HSAD).

HLAD has been studied in detail. The mode of reduction follows Prelog's rule, and a model of the active site of the enzyme has been proposed, giving favoured and forbidden regions for a prospective substrate (Figure 4.7). Thus for a 2(*S*)-alkylcyclohexanone, the *trans*-product (6) is obtained, while the 2(*R*)-compound is unreactive. In contrast, a racemic mixture of the thiaketone (7) gives two diastereoisomeric alcohols by reduction of the enantiomers through the "best-fit" conformations (Scheme 4.10).

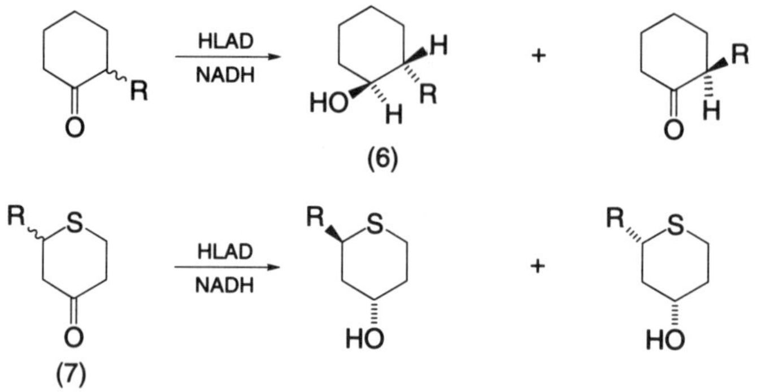

Scheme 4.10.

TBAD displays a similar selectivity on substrates other than very simple ketones. Like HLAD, this enzyme can exhibit enantioselectivity in the reduction of racemic ketones, where one enantiomer cannot fit into the active site of the enzyme (Scheme 4.11). Thus the ketone (8) yields the optically active alcohol (9) and the recovered, optically active ketone. As expected, TBAD is not able to reduce the sterically congested ketones

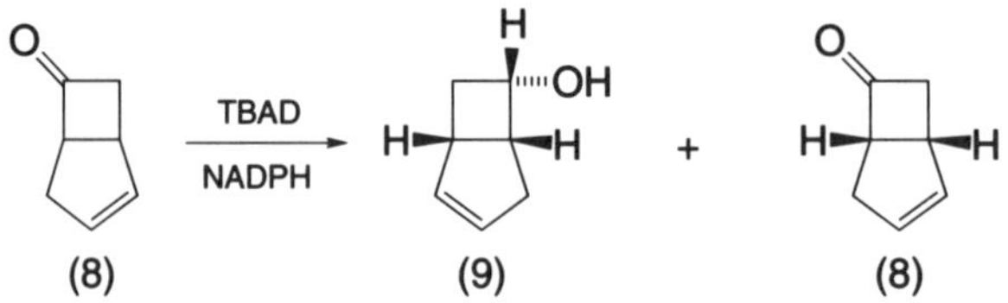

Scheme 4.11.

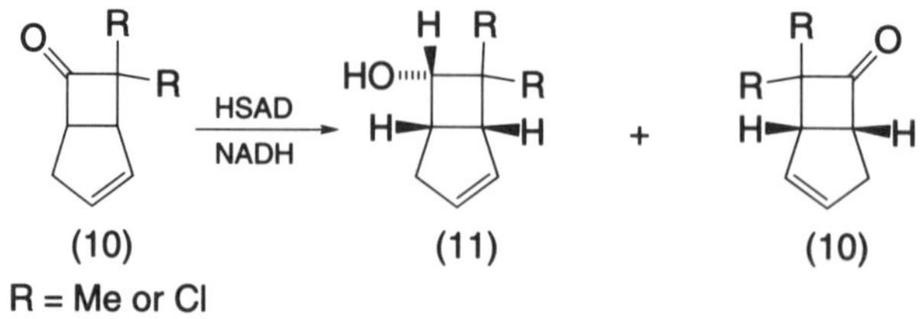

R = Me or Cl

Scheme 4.12.

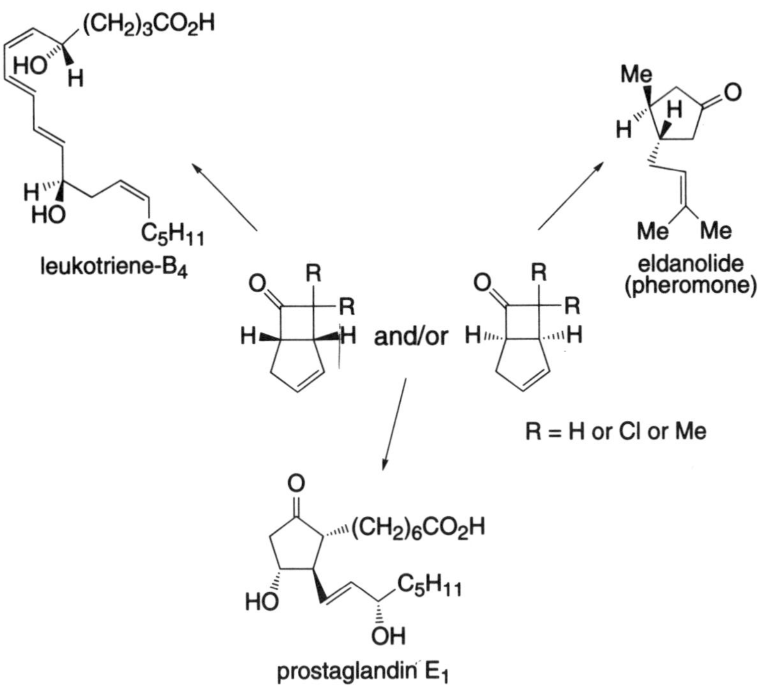

Figure 4.8. Use of bicyclo (3.2.0.)hept-2-en-6-ones as synthetic intermediates.

(10). However, HSAD catalyses the reduction, giving the alcohols (11) (>95% e.e.) [note the (S)-configuration at the newly formed chiral centre (Scheme 4.12)].

The importance of the resolution of bicyclic compounds of types (9) and (11) is due to the fact that they can be transformed by multistage conventional chemical procedures into biologically active compounds (such as prostaglandins and leukotrienes) and pheromones (such as eldanolide) which may be of future use in pest control (Figure 4.8).

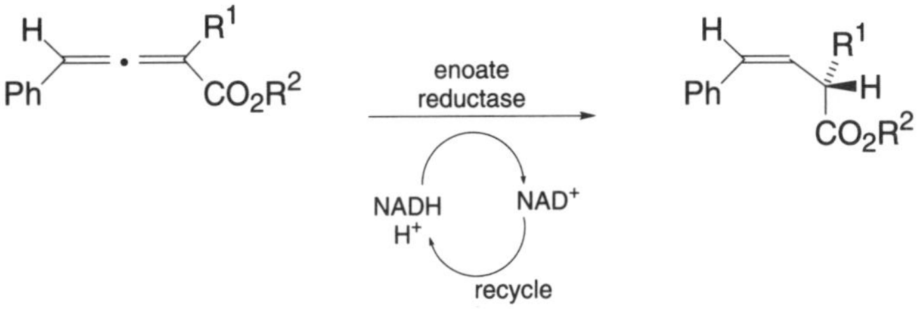

Scheme 4.13.

Other enzymes, the enoate reductases, reduce α,β-unsaturated esters, aldehydes, and so forth. These enzymes have been isolated, and some impressive transformations have been recorded (e.g. Scheme 4.13). However, co-factor regeneration and enzyme stability can be problematic, and this sort of biotransformation is not yet at a stage at which it can be used on a substantial scale by a non-expert.

4.5 Oxidation of alcohols to aldehydes or ketones

The enzyme-catalysed oxidation of alcohols to carbonyl compounds is not as attractive as the reverse reaction discussed earlier. This is because the oxidation often removes a chiral centre from the substrate, and the recycling of oxidized co-factors $NAD(P)^+$ can be problematic. However, there are some instances where the enzyme technology has an advantage over conventional chemical oxidants. For example, the polyol D-sorbitol is oxidised by the micro-organism *Acetobacter suboxydans* to give L-sorbose (Scheme 4.14) (see also Chapter 1, Section 1.7).

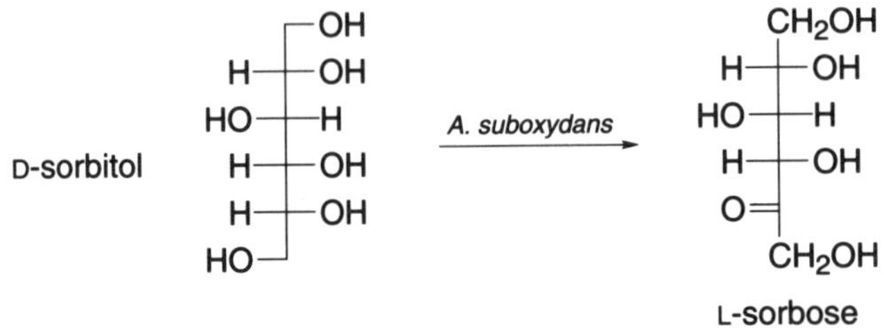

Scheme 4.14.

Isolated enzymes can also be used to catalyse selective oxidation reactions. For, example, galactose oxidase oxidizes xylitol into (L)-xylose (Scheme 4.15), while HLAD catalyses the stereoselective oxidation of prochiral diols of the type (12) to give, initially, a chiral hydroxyaldehyde

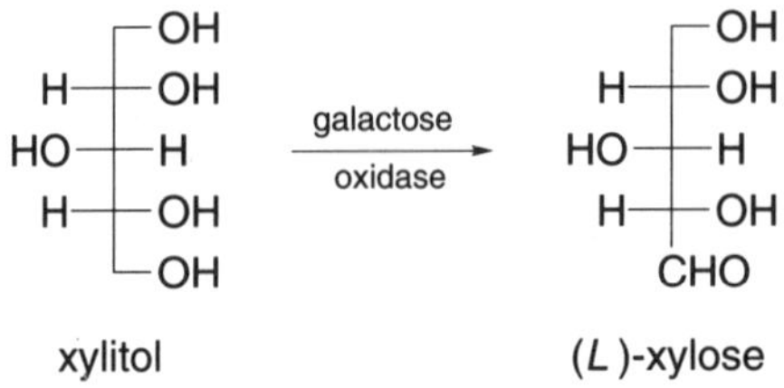

Scheme 4.15.

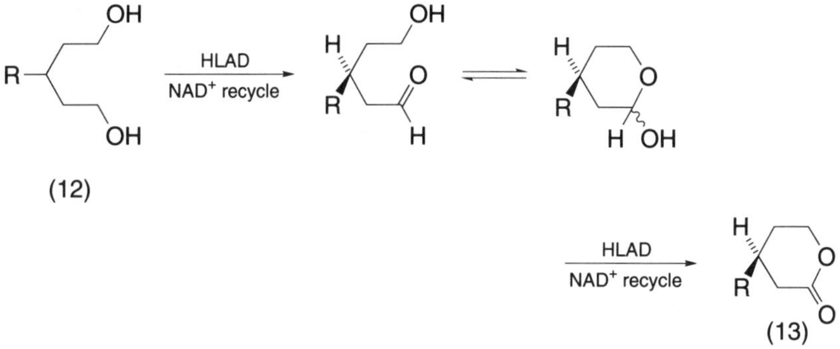

Scheme 4.16.

which undergoes spontaneous cyclization and further oxidation by the dehydrogenase to furnish lactones (13) as the final products (Scheme 4.16). In a similar way, *meso*-diols can give good yields of optically pure lactones. One example is given in Scheme 4.17. Once again it must be said that recycling procedures for NAD$^+$ are not straightforward.

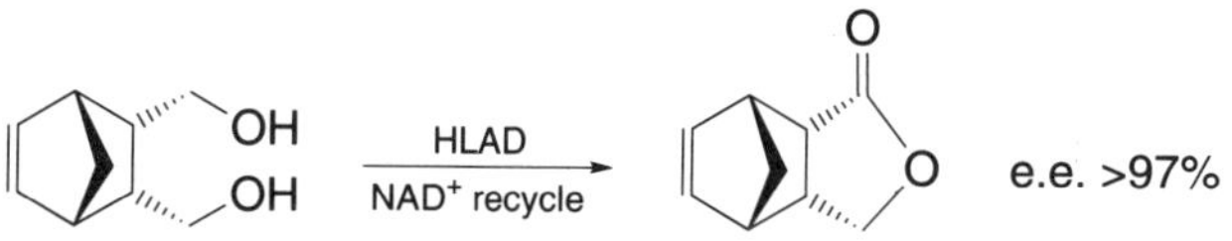

Scheme 4.17.

4.6 Oxidation of ketones to esters or lactones

The conversion of an acyclic ketone into an ester or a cyclic ketone into a lactone is known as the Baeyer-Villiger reaction and can be accomplished chemically using a peracid (Scheme 4.18). Certain advantages can be

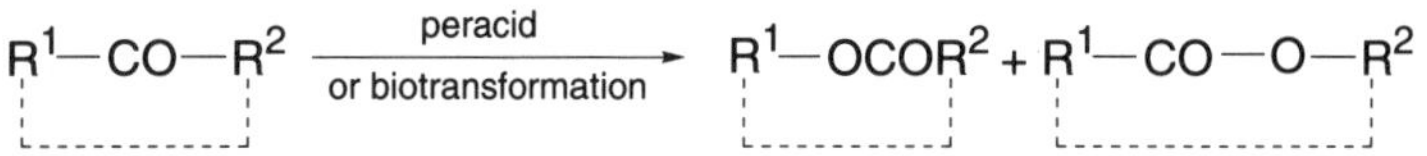

$$R^1-CO-R^2 \xrightarrow[\text{or biotransformation}]{\text{peracid}} R^1-OCOR^2 + R^1-CO-O-R^2$$

Scheme 4.18.

obtained by using mono-oxygenase enzymes for the same task. For example, the regioselectivity of the reaction can be altered, and optically active materials can be obtained. This bioconversion has received a good deal of attention, and some of the finer points of the process have been discussed in Chapter 2. In a direct comparison of the chemical and biological methodologies it was noted that peracid oxidation of the cyclobutanone derivative (8) gave a mixture of the racemic lactones (14) and (15) in the ratio 9:1. In contrast, oxidation of the micro-organism *Acinetobacter calcoaceticus* gives equal amounts of the two lactones, each in an optically pure form (Scheme 4.19). In other instances the same organism gives rise

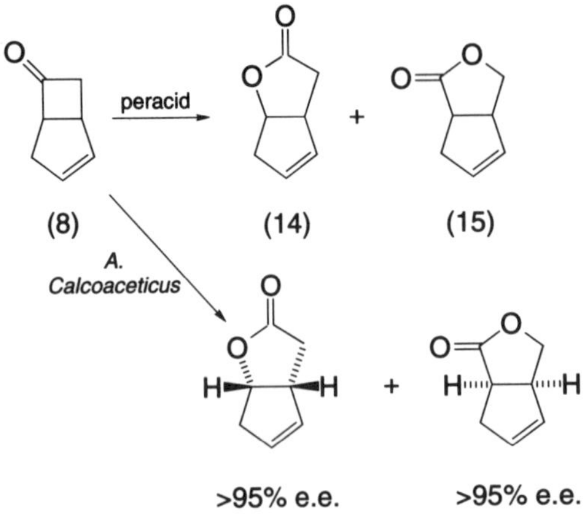

Scheme 4.19.

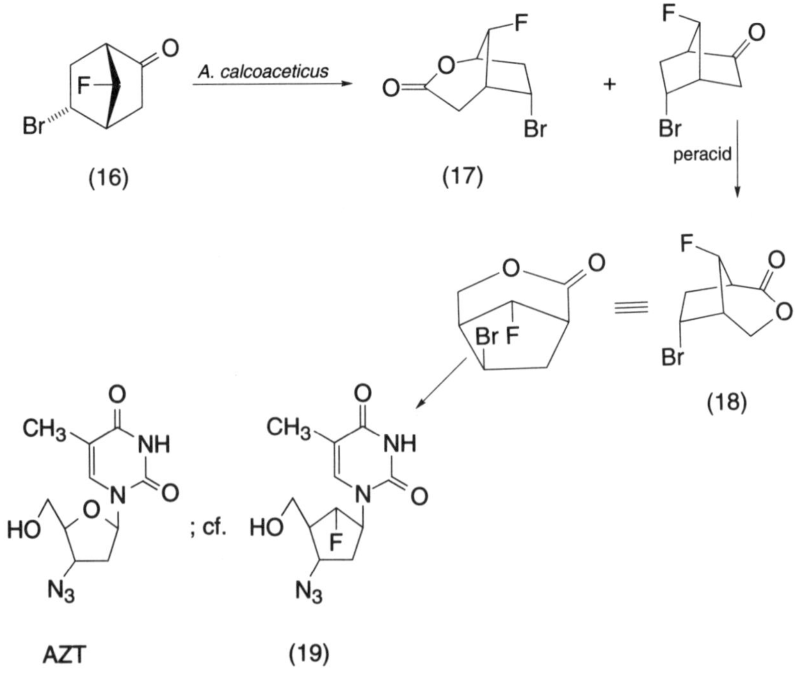

Scheme 4.20.

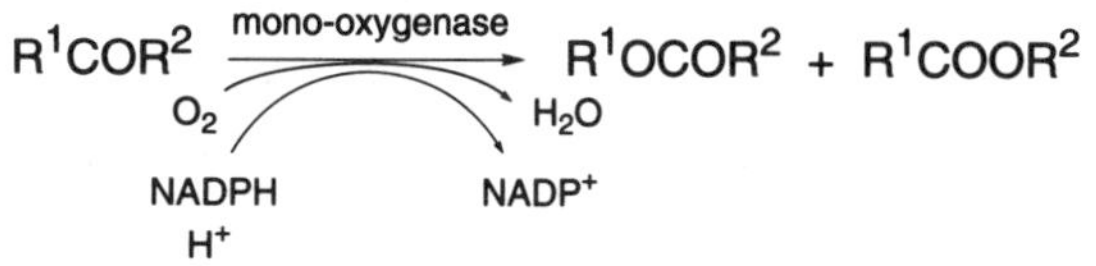

$$R^1COR^2 \xrightarrow{\text{mono-oxygenase}} R^1OCOR^2 + R^1COOR^2$$

O_2 H_2O
NADPH NADP$^+$
H$^+$

Figure 4.9. Mono-oxygenase-catalysed Baeyer-Villiger reactions.

to an enantioselective oxidation. As shown in Scheme 4.20, the dihalo-ketone (16) is oxidized by *A. calcoaceticus* to give the lactone (17) and the recovered, optically pure ketone. The recovered ketone was oxidized by peracid to give the lactone (18) in good yield (once again note the different regioselectivity of the oxidation). The lactone (18) was used to prepare the anti-HIV agent (19), a close analogue of the clinically important nucleoside analogue AZT.

As discussed in some detail in Chapter 2, the use of a micro-organism for this reaction can be fraught with difficulties, for example through further biotransformation of the lactone. The alternative is to use the isolated mono-oxygenase enzyme which must be reconstituted with NADPH (the requisite co-factor) for the reaction to commence (Figure 4.9). The necessity of recycling the co-factor is the drawback to this approach. To circumvent this problem, it has been shown that a dehydrogenase enzyme and a mono-oxygenase enzyme can work in concert, with in situ recycling of the co-factor, to provide a "user-friendly" method for the conversion of a secondary alcohol into the corresponding lactone (Scheme 4.21).

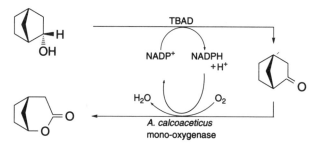

Scheme 4.21.

4.7 Hydroxylation of aliphatic and aromatic compounds

The microbial hydroxylation of aliphatic compounds can effect the functionalization of molecules at positions distant from pre-existing groups. Such functionalizaton of non-activated positions is difficult to emulate using conventional chemistry. A good deal of research in this area has concerned the functionalization of steroids, and a portfolio of micro-organisms is available that is capable of selectively oxidizing almost any position on the steroid nucleus. This research has been rewarded by the discovery of a method for the hydroxylation of the 11-position of the tetracyclic system. That has achieved the conversion of progesterone (20) into 11-hydroxy-progesterone (21), obviated a tedious chemical route, and given access to highly desirable anti-inflammatory agents for use in the clinic (Scheme 4.22) (for a broader discussion, see Section 6.3).

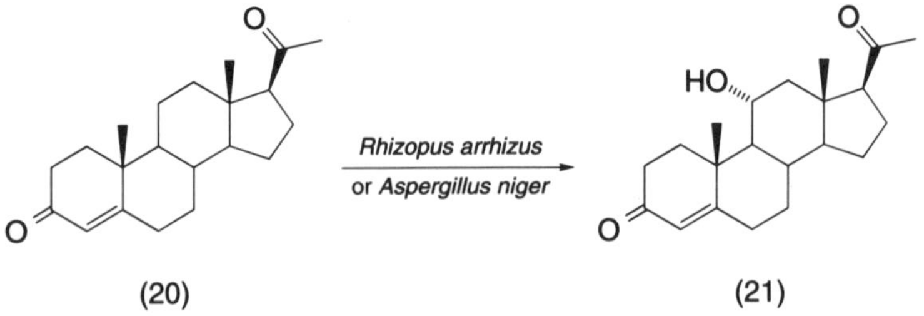

Rhizopus arrhizus
or *Aspergillus niger*

(20)   (21)

Scheme 4.22.

A lot of other studies have focussed on other types of molecules, such as naturally occurring terpenes and alkaloids, as well as unnatural alicyclic and heterocyclic compounds. Yields of selectively hydroxylated compounds can be good to very good, as indicated in Scheme 4.23. The micro-

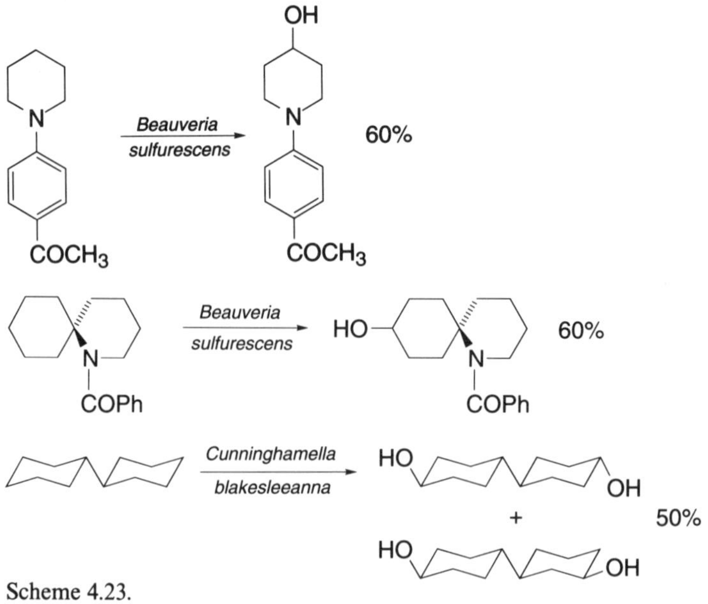

*Beauveria
sulfurescens*   60%

*Beauveria
sulfurescens*   60%

*Cunninghamella
blakesleeanna*

+   50%

Scheme 4.23.

organism *Beauveria sulfurescens* has been studied in depth, but even after concentrating much research on this single organism it is still difficult to predict the outcome of the hydroxylation on a new substrate. Despite the obvious successes in the field, more research is needed before such hydroxylations, remote from pre-existing functionality, become useful to the synthetic organic chemist.

Hydroxylation reactions adjacent to electron-rich units have proved to be very useful. For example, benzylic hydroxylation of the tetrahydroquinoline derivative (22) gives oxamniquine, an anthelmintic (worm-control agent) used in Africa and South America against human parasites (Scheme 4.24).

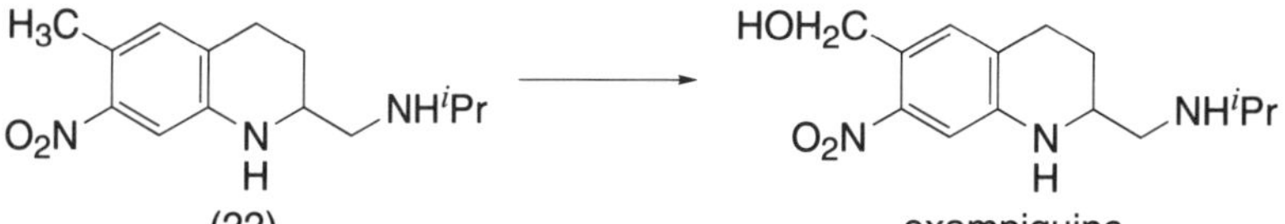

Scheme 4.24.

Hydroxylation of aromatic compounds furnishing phenols is a process that has found favour in some instances, for example in the preparation of prenalterol, a compound with pronounced pharmacological activity (Scheme 4.25). However, the transformation of aromatic compounds that

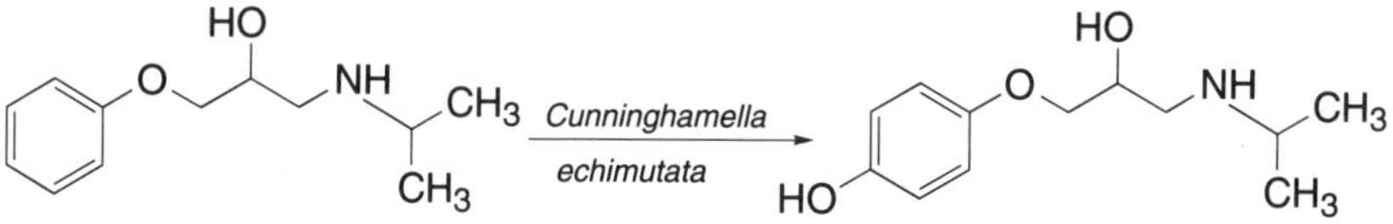

Scheme 4.25.

really catches the eye is shown in Scheme 4.26. This is because there is no equivalent chemical process, and the non-aromatic cyclohexadienediols make excellent starting points for organic synthesis, as discussed later. The conversion of benzene into cyclohexadienediol was observed in the early 1970s. A group of scientists at Imperial Chemical Industries (ICI) developed the process some 10 years later, a useful *Pseudomonas* being found under a benzene storage tank. The organism had utilized benzene as an energy source via oxidation to the dienediol, formation of catechol and further oxidation to give small fragments. Mutation to remove the

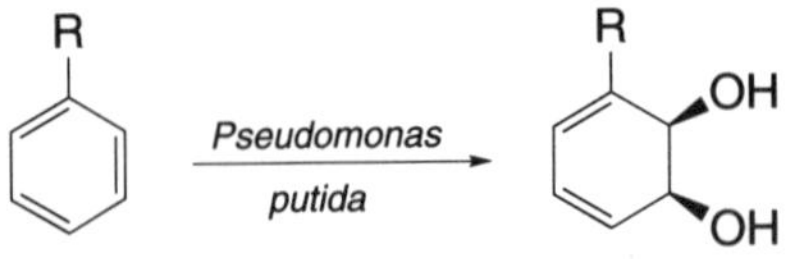

Scheme 4.26.

enzyme responsible for the conversion of the dienediol to catechol gave
an organism which overproduced the desired non-aromatic compound.

Cyclohexadienediol in its protected form (23) has been used to prepare
some interesting natural products. (Figure 4.10). A bonus is obtained when
mono-substituted benzenes are fed to the organism, since, in the majority
of cases, optically active diols are obtained possessing the absolute
configuration detailed in Scheme 4.26. For example, the diol derived from
chlorobenzene has been used to make synthetic building blocks (synthons)
for pyrrolizidine alkaloids and sugars such as L-ribonolactone (Figure 4.11).

This area of work serves to show that judicious manipulation of the
metabolic pathway of an organism can lead to immensely useful materials
for synthesis. Thus, instead of simply making use of the secondary
metabolites of the organisms (compounds in the chiral pool), we can
encourage the organisms to make more versatile "designer" building
blocks for organic chemists.

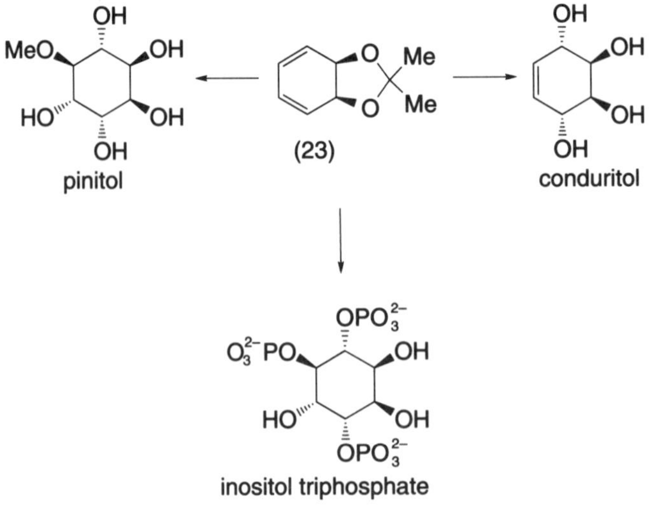

Figure 4.10. Use of cyclohexadienediol derivative in organic synthesis.

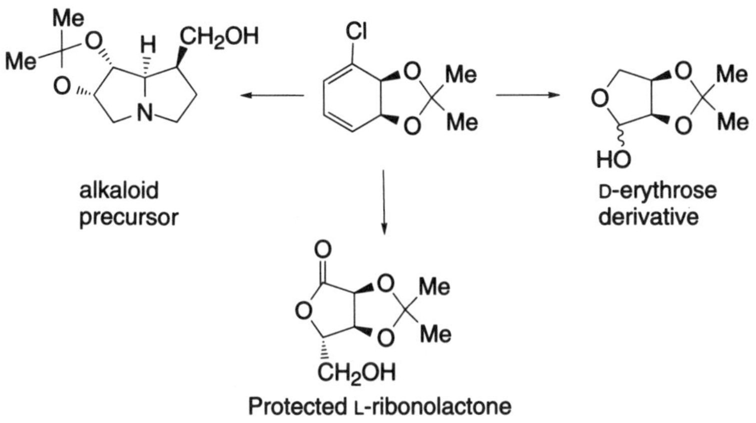

Figure 4.11. Use of a chlorocyclohexadienediol derivative in organic synthesis.

4.8 Oxidation of alkenes and sulphides

In contrast to the developments in the areas of arene oxidation, a general method for the controlled epoxidation of alkenes has not been found. There have been reports that indicate that terminal alkenes can be converted into optically active epoxides, but there is nothing yet to rival the Sharpless oxidation of allylic alcohols using a chiral titanium catalyst (Scheme 4.27) as a general synthetic method. However, it should be

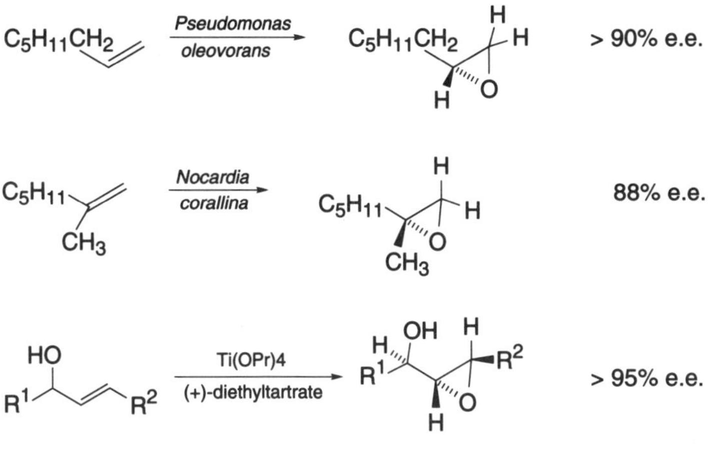

Scheme 4.27.

mentioned that fosfomycin (24) was synthesized in optically active form from the corresponding alkene using *Penicillium spinulosum* (Scheme 4.28),

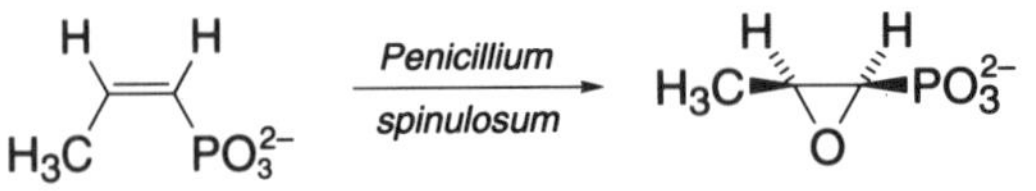

(24)

Scheme 4.28.

whereas classical organic chemistry did not provide a simple method for preparation of the compound.

The conversion of sulphides into sulphoxides is another transformation that takes a non-chiral compound to a chiral entity. Once again the oxidation can be accomplished using a whole-cell system or an enzyme. Aryl alkyl sulphides are oxidised by *Corynebacterium equi* to give (R)-sulphoxides in good to excellent optical purity (Scheme 4.29). Yields can be variable, one problem being overoxidation of the sulphoxide to the corresponding sulphone. This problem is circumvented by the use of an isolated enzyme, horseradish peroxidase, but unfortunately the optical purities of the products are lower.

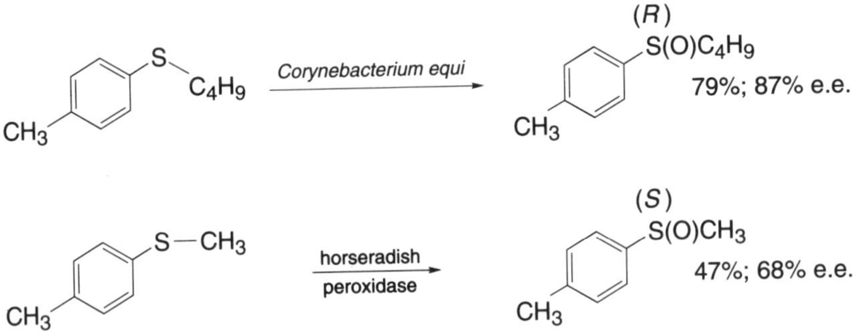

Scheme 4.29.

The sulphoxides so formed are useful in synthetic organic chemistry, since, through formation and reaction of the corresponding α-carbanion (e.g. with aldehydes), new chiral centres can be induced adjacent to the sulphoxide unit.

4.9 Conclusions and overview

Enzyme-catalysed or whole-cell-mediated oxidation and reduction reactions are generally less easy to perform than, for example, hydrolysis reactions. There are exceptions to this rule, for example the yeast reaction of ketones, and some processes have attained commercial significance, for example the modification of progesterone, the preparation of vitamin E intermediates and the synthesis of oxamniquine.

However, the lack of predictability of some biotransformations, such as hydroxylation of methine, methylene or methyl groups remote from pre-existing functionality, means that there must be several more years of research work before the possible emergence of a general methodology. A selection of recent references is to be found in the Bibliography, and this list is the suggested starting point for gathering further information.

Bibliography

Czsuk, C., & Glänzer, B. I. (1991). Baker's Yeast Mediated Transformations in Organic Chemistry. *Chem. Rev.*, *91*, 49.

Fuganti, C. (1990). Baker's Yeast Mediated Synthesis of Natural Products. *Pure Appl. Chem.*, *62*, 1449.

Holland, H. L. (1992). *Organic Synthesis with Oxidative Enzymes*. VCH Publishers, Weinheim.

Hummel, W., & Kula, M. R. (1989). Dehydrogenases for the Synthesis of Chiral Compounds. *Eur. J. Biochem.*, *184*, 1.

Servi, S. (1990). Baker's Yeast as a Reagent in Organic Synthesis. *Synthesis*, 1.

Walsh, C. T., & Chen, Y. C. (1988). Enzyme Baeyer-Villiger Oxidations by Flavin Dependant Mono-oxygenases. *Angew. Chem. Int. Ed.*, *27*, 333.

Wood, O. P., & Young, C. S. (1990). Reductive Biotransformations of Organic Compounds by Cells or Enzymes of Yeast. *Enzyme Microb. Technol.*, *12*, 482.

Yamada, H., & Shimizu, S. (1988). Microbial and Enzymatic Processes for the Production of Biologically and Chemically Useful Compounds. *Angew. Chem. Int. Ed.*, *27*, 622.

5

Useful intermediates and end-products obtained from biocatalysed carbon–carbon, carbon–oxygen, carbon–nitrogen and carbon–chalcogen bond-forming reactions

5.1 Introduction

This chapter deals with three important classes of biotransformations. Firstly, those enzymes that catalyse the stereoselective formation of carbon–carbon bonds will be examined. These enzymes, whose natural functions often are to degrade carbohydrate-like molecules, have proved to be versatile catalysts for C—C bond synthesis. Secondly, we shall look at those enzymes that mediate the formation of C—X bonds, where X = O, N, S, Hal (halogen). These enzymes are termed lyases (see Table 2.1) and often carry out very simple reactions (e.g. the addition of water to a double bond) with very high stereoselectivity and regioselectivity. Finally, the application of a range of enzymes (including C—C bond formation) to carbohydrate synthesis will be examined. This chapter will conclude with some examples of the ways in which multienzyme reactions can be constructed to enable highly complex molecules to be assembled in an efficient manner.

5.2 Enzyme-catalysed asymmetric carbon–carbon bond formation

5.2.1 Aldolases

The synthesis of carbon–carbon bonds, particularly with control of stereochemistry, has been an area of intense activity in recent years. Much attention has been focussed on the asymmetric aldol reaction, principally using stoichiometric quantities of a chiral auxiliary, but more recently in the catalytic mode. The use of aldolases for the asymmetric construction of carbon–carbon bonds presents a potentially useful complementary methodology.

The aldolases are a diverse class of enzymes that catalyse the coupling of a carbonyl-containing compound (nucleophile), containing one, two or three carbons, with an aldehyde (electrophile). In most cases the nucleophile is either pyruvic acid or dihydroxyacetone phosphate, whereas the electrophilic aldehyde is much more variable in structure. In many cases the reaction generates two new stereogenic centres in the product. In general, only one isomer is obtained from the four possible stereoisomeric products (Scheme 5.1).

Scheme 5.1.

From a synthetic point of view, aldolases offer a number of advantages as catalysts for C—C bond formation. For example, they operate best on unprotected substrates, thus avoiding the problem of complex protection/deprotection schemes for polyfunctional molecules (e.g. carbohydrates). They also catalyse C—C bond synthesis with high diastereoselectivity and enantioselectivity. Such simultaneous control is often difficult to achieve using non-enzymic aldol reactions.

To date, the best-studied aldolase has been fructose-1,6-diphosphate aldolase from rabbit muscle. Rabbit muscle aldolase (RAMA) catalyses the condensation of D-glyceraldehyde-3-phosphate (1) with dihydroxyacetone phosphate (2) to yield fructose-1,6-diphosphate (3) (Scheme 5.2).

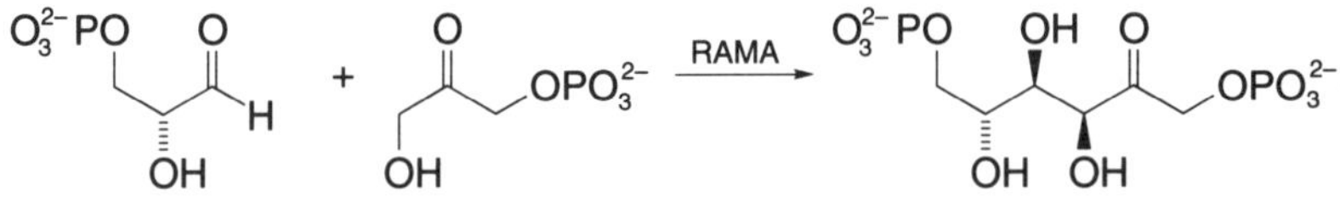

Scheme 5.2.

RAMA has a high specificity for dihydroxyacetone phosphate (DHAP), but a low specificity for the electrophilic aldehyde, to the extent that more than 75 aldehydes can be coupled. Most importantly, all the products have the same absolute configuration (*threo* and D-*glycero*) about the newly formed C3–C4 bond (Scheme 5.3). Although RAMA is limited to

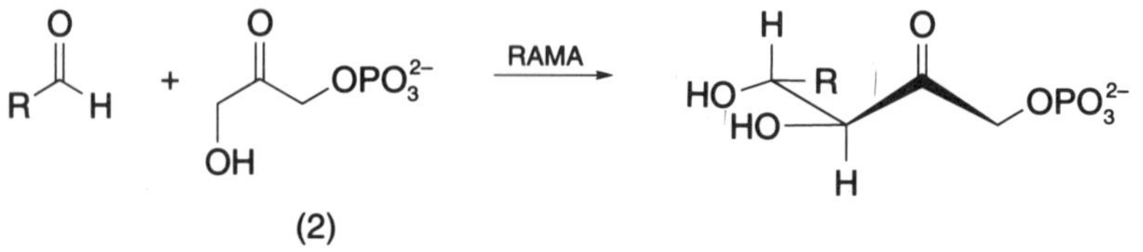

(2)

Scheme 5.3.

the production of the *threo* stereochemistry, aldolases that can generate
the remaining three vicinal diol stereochemistries are becoming available.
They all use DHAP as one component of the condensation, but differ in
the electrophile used (Scheme 5.4).

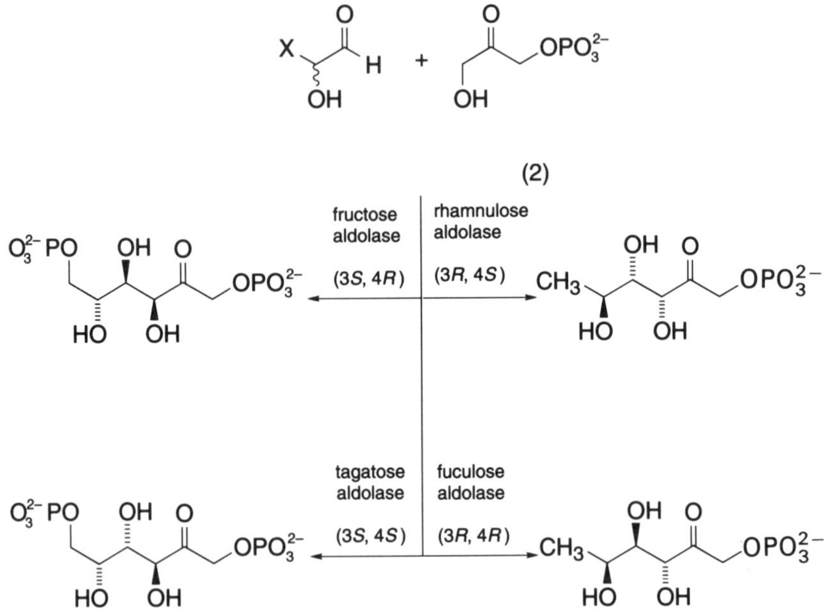

Scheme 5.4.

RAMA has been used to catalyse the key C—C bond-forming step in the
synthesis of a number of natural products [e.g. deoxynojirimycin (4)
(Scheme 5.5)]. Note that the enzyme phosphatase is used to remove the
phosphate group under mild conditions.

Since many of the aldolases use DHAP as the nucleophilic component
of the reaction, attention has been given to devising efficient methods for
its preparation starting from dihydroxyacetone. The synthesis of DHAP
can be achieved in high overall yield according to the procedure shown

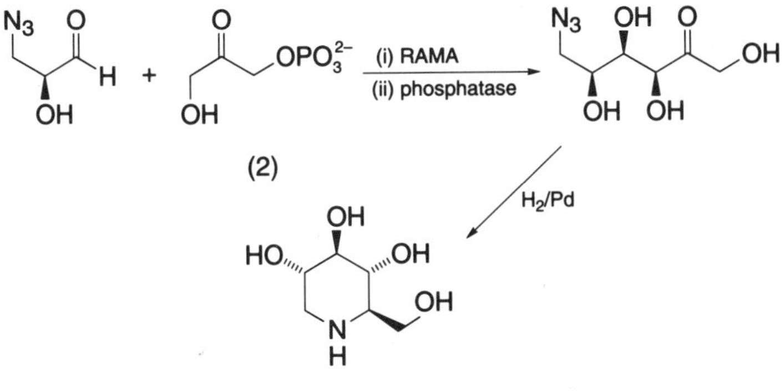

(4)

Scheme 5.5.

in Scheme 5.6. In this synthesis, dihydroxyacetone is initially converted to a protected dimer in which only one of the original hydroxyl groups is now available for reaction with the phosphorylating reagent phosphorus oxychloride. The phosphorylated dimer can then be cleaved under aqueous acidic conditions to yield DHAP (2).

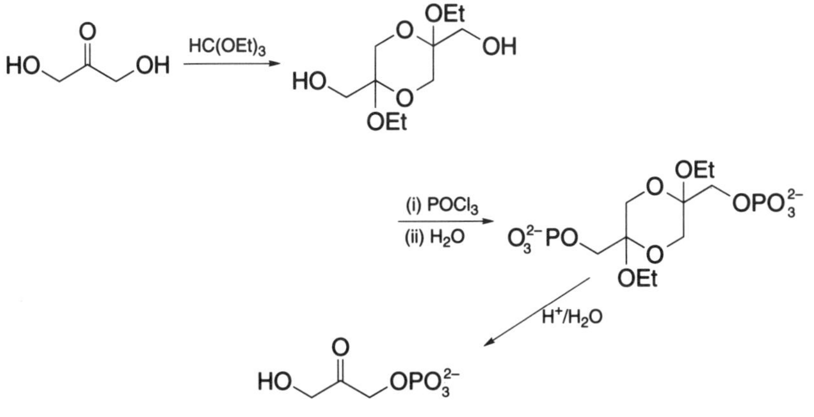

(2)

Scheme 5.6.

An ingenious alternative to the use of DHAP is to exploit the fact that organic arsenates can be spontaneously formed in situ from the corresponding alcohol and inorganic arsenate ($HOAsO_3^{2-}$). Thus dihydroxyacetone arsenate (5) can be prepared in solution, and it has been shown to be an effective mimic of DHAP, enabling coupling with an aldehyde to take place. The product then hydrolyses to release inorganic arsenate, thus completing the cycle (Scheme 5.7).

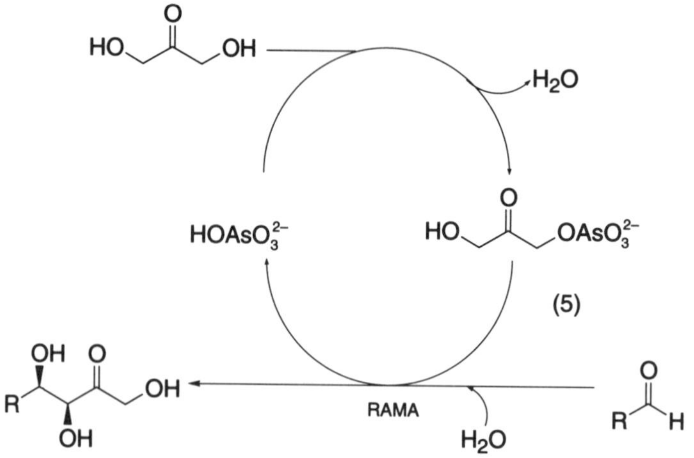

Scheme 5.7.

A second example of the use of RAMA in natural-product synthesis is given by the synthesis of 3-deoxy-D-*arabino*-heptulosonic acid 7-phosphate (6), an important intermediate in the biosynthesis of aromatic amino acids. Again RAMA catalyses the important stereospecific C—C bond-forming step, but this time there is an additional advantage in that the phosphate group, which is required in the target molecule, is introduced concurrently with C—C bond formation (Scheme 5.8).

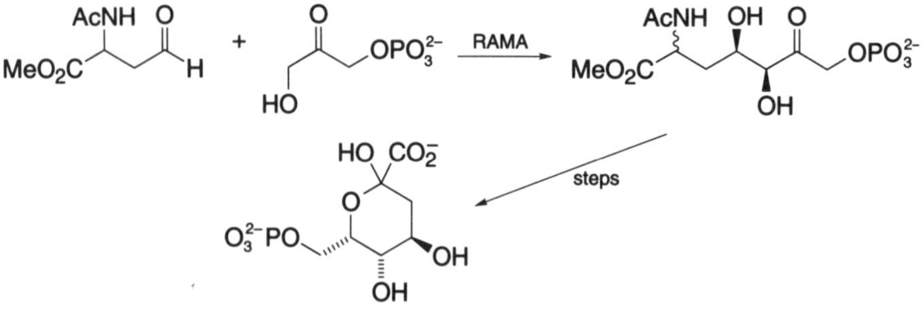

(6)

Scheme 5.8.

Recently, other aldolases have been used, such as sialic acid (*N*-acetylneuraminic acid) aldolase, which catalyses the condensation between pyruvate as the nucleophile and the open-chain form of *N*-acetylmanno-samine (7) as the electrophilic component of the reaction (Scheme 5.9). In

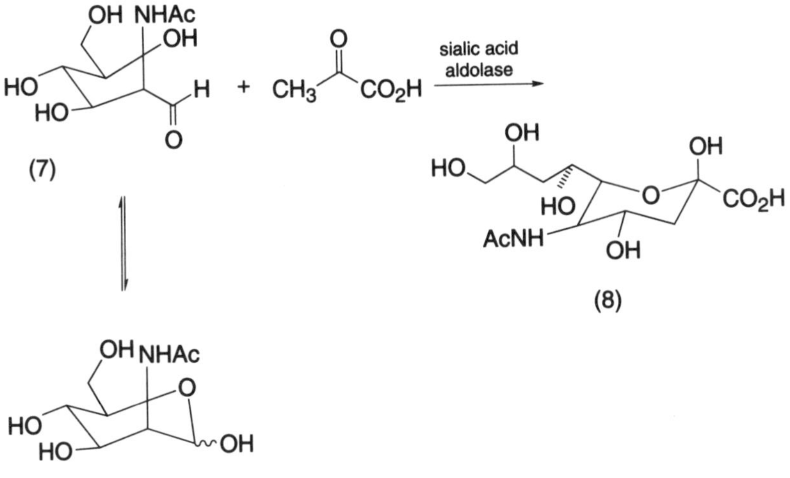

Scheme 5.9.

this case, only one new stereogenic centre is formed, but again with complete control of stereochemistry. Sialic acid (8) is the most important member of the sialic acids, which themselves are constituents of the oligosaccharide portions of glycopeptides and glycolipids. This aldolase is commercially available (from *Clostridium perfringens*) and is produced on an industrial scale using a gene-cloned strain of *E. coli*. In a similar fashion to RAMA, sialic acid aldolase is highly specific for the nucleophile (pyruvate), but will accept a range of analogues of *N*-acetylmannosamine as the electrophile, thus allowing the preparation of a range of structural analogues of sialic acid. In view of the lengthy and complex syntheses required for the sialic acids, the aldolase approach is potentially an efficient way of preparing this class of compounds.

A final example of an aldolase-catalysed reaction is that between D-glyceraldehyde-3-phosphate (1) (electrophile) and acetaldehyde (nucleo-phile) (Scheme 5.10). It is interesting to note that the enzyme carefully organizes the reactants such that acetaldehyde acts only as the nucleophile component of the reactant, but not the electrophile. This aldolase has been shown to feature relaxed specificity with respect to both of the substrates.

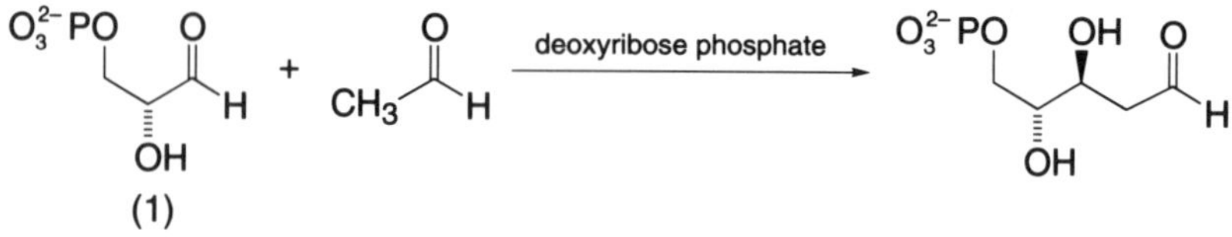

(1)

Scheme 5.10.

5.2.2 Other C—C bond-forming enzymes

An enzyme closely related to the aldolases is transketolase. The enzyme is commercially available (from baker's yeast) and can also be obtained from spinach leaves. Transketolase catalyses the stereospecific synthesis of C—C bonds using aldehydes as the electrophiles, with a suitable 2-carbon ketol donor [e.g. hydroxypyruvate (9)] as the nucleophile (Scheme 5.11). The use of hydroxypyruvate ensures that the reaction goes to completion, since carbon dioxide is evolved as the by-product, and hence the reaction is irreversible. In addition, both magnesium ions and catalytic thiamine pyrophosphate are required as co-factors.

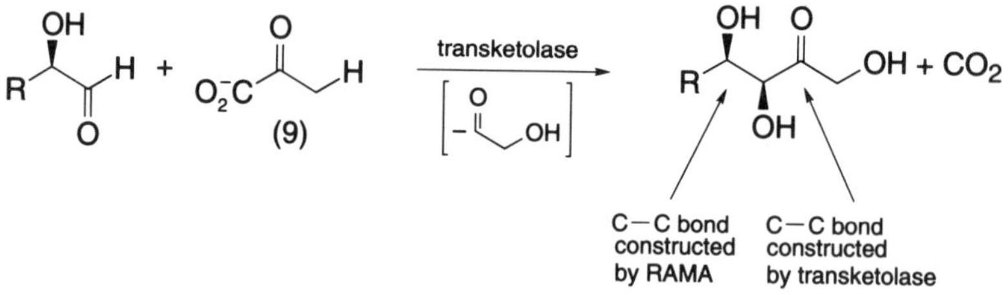

Scheme 5.11.

The products obtained from the transketolase reaction have the same D-*threo* stereochemistry as the products derived from the RAMA-catalysed reactions. However, as indicated in Scheme 5.11, the carbon–carbon bond that is constructed differs. As might be predicted, transketolase is highly specific for hydroxypyruvate, but will accept a range of aldehydes. Transketolase has recently become more readily available owing to overexpression of the gene in *E. coli* (cf. sialic acid aldolase). Its application to natural-product synthesis has recently been demonstrated by the synthesis of (+)-*exo*-brevicomin (10) (Scheme 5.12). In this synthesis, only one of the enantiomers of the starting aldehyde reacts with hydroxypyruvate, giving a product that is a single diastereomer.

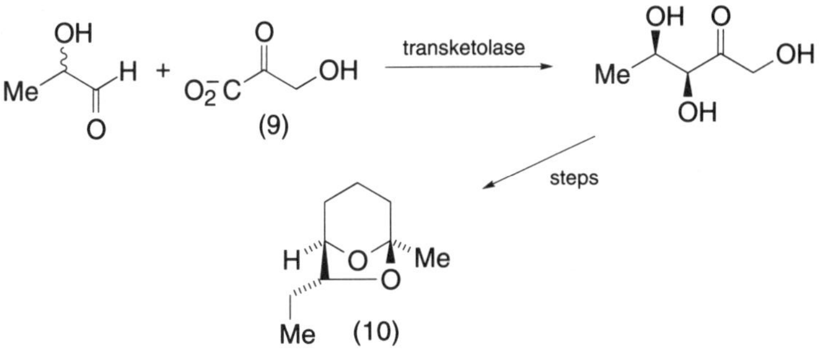

Scheme 5.12.

Mandelonitrile lyase belongs to the oxynitrilase group of enzymes and is obtainable in large quantities from sweet almonds. It catalyses the synthesis of (*R*)-cyanohydrins from the corresponding aldehyde (or ketone) and HCN (Scheme 5.13). Cyanohydrins are versatile starting materials for the synthesis of a range of chiral intermediates, including α-hydroxy acids and β-hydroxy amines. If the reaction is carried out with water as the solvent, then the optical activities of the products are found to be modest, owing to two factors, namely (i) the competing non-enzyme-catalysed addition of cyanide ion to the carbonyl group and (ii) racemisation of the product due to reversal of the reaction. However, by switching to a low-water system (e.g. 95% EtOAc) as the solvent, both of these factors are greatly suppressed, leading to products with excellent optical purities. An alternative solution is to use acetone cyanohydrin as the source of cyanide instead of NaCN/KCN. Acetone cyanohydrin generates cyanide in situ, but at much lower concentrations, and hence the rate of the non-enzyme-catalysed cyanohydrin formation is greatly reduced. Recently, (*S*)-oxynitrilases have been isolated (e.g. from millet). However, they appear to be more restricted in the range of substrates they can use.

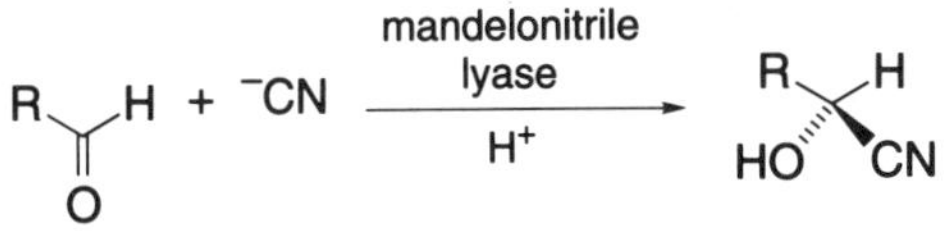

Scheme 5.13

Baker's yeast has long been known to catalyse the synthesis of carbon–carbon bonds using acyloin-type processes. Thus, by using α,β-unsaturated aldehydes with fermenting yeast containing ethanol, it is possible to obtain products in which a 2-carbon unit (derived from ethanol) has been added in a stereospecific manner (Scheme 5.14). The reactions can be carried out on a reasonable scale (up to 10 g) and proceed in modest (25%) overall yield. The reactions are known to proceed via the ketone, and indeed the ketone can be isolated by the addition of acetaldehyde. An interesting application of this reaction is shown by the synthesis of α-tocopherol (vitamin E) (14) in Scheme 5.15. Using the furan-containing aldehyde (11),

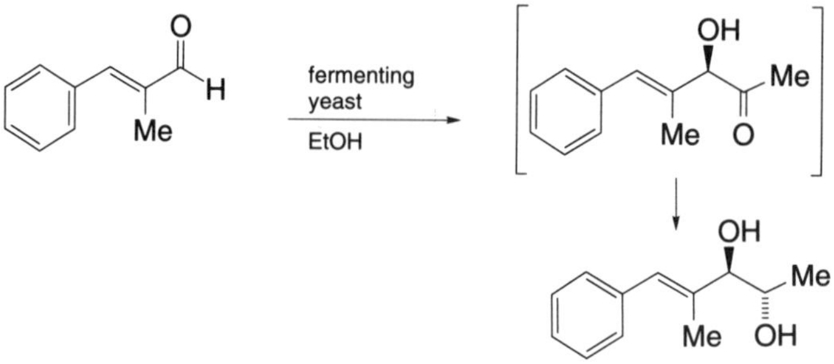

Scheme 5.14.

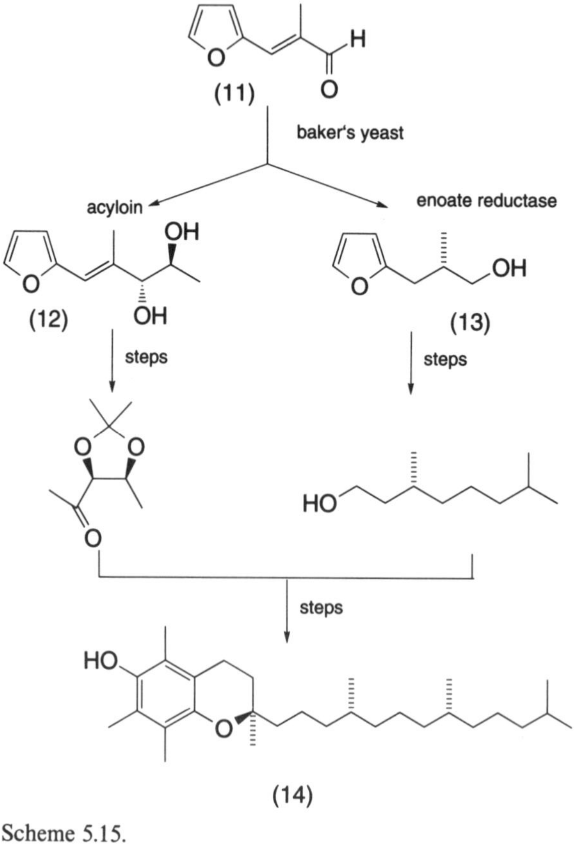

Scheme 5.15.

two products were obtained from baker's yeast, namely the expected acyloin product (12) (in 20% yield) and the saturated compound (13) resulting from an enoate reductase activity.

A related reaction catalysed by baker's yeast is shown in Scheme 5.16. By using cyanoacetone as the substrate, again a stereospecific 2-carbon addition occurs, giving both the *syn* and *anti* products (15) (each >99% e.e.) in a combined yield of 88%.

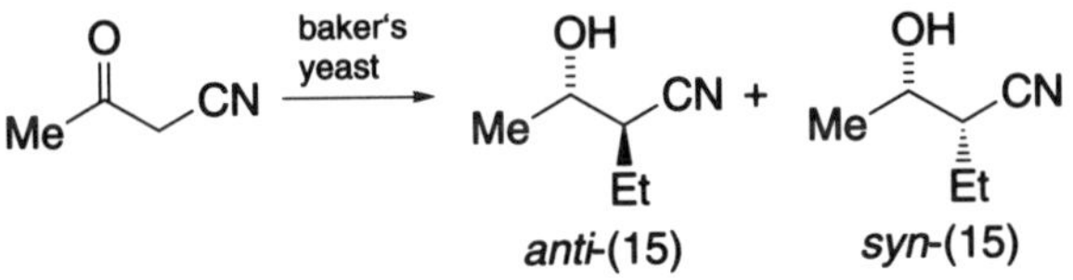

Scheme 5.16.

Michael-type reactions have been achieved with baker's yeast using trifluoroethanol as the co-substrate (Scheme 5.17). These transformations probably are related to the acyloin condensation in that it has been proposed that the trifluoroethanol is initially oxidized enzymically to the aldehyde, which is then converted to a reactive "acyl anion" that can add to the electrophilic enone in a conjugate manner. The α,β-unsaturated ketones are good Michael acceptors, leading to the corresponding γ-hydroxyketones (via the diketone intermediate) in moderate yields (26–40%) and good optical purities (e.e. =79%). The absolute configurations of the products (16) were not determined.

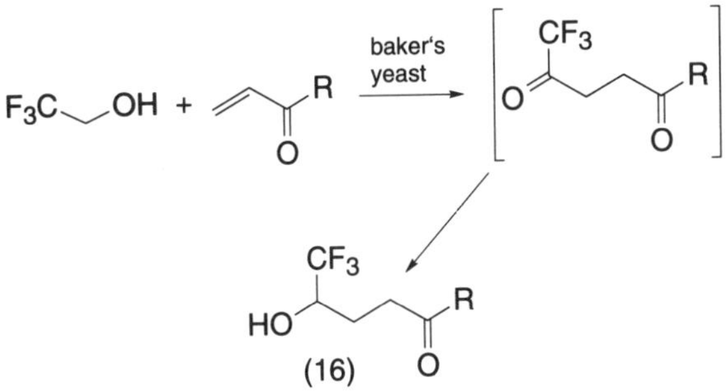

Scheme 5.17.

Finally, an interesting and potentially useful baker's-yeast-mediated reaction has been demonstrated with the conversion of squalene oxide (17) to lanosterol (18) (Scheme 5.18). Two facets of this reaction are noteworthy: Firstly, if racemic squalene oxide is used, then the lanosterol

that is obtained is optically pure (in 80% yield). Secondly, it is essential
that the yeast be ultrasonicated prior to use. The latter observation may
imply that the enzyme responsible for this cyclization is membrane-located.

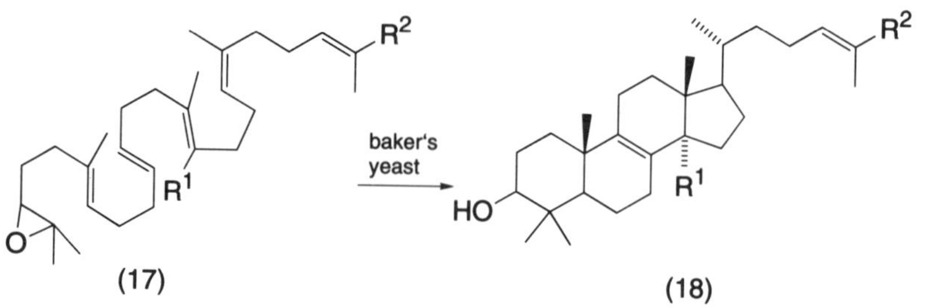

Scheme 5.18.

5.3 Synthesis of C—X bonds (where $X = $O, N, S, Hal) using lyases and haloperoxidases

Enzymes capable of effecting C—O and C—N bond synthesis have also
proved useful. Typically these reactions occur with high regioselectivity
and involve stereospecific *trans* or *anti* addition of H—X to the double
bond. For example, fumarase, which naturally catalyses the addition of
water to fumaric acid (19, $X = $H), yielding malic acid (n.b. this reaction is
performed on a large scale yielding more than 40 tonnes per month), also
catalyses the analogous reaction using chlorofumaric acid (19, $X = $Cl) as
the substrate (Scheme 5.19) (see Section 6.4.4). The product L-*threo*-
chloromalic acid (20) is obtained in optically pure form and can be
elaborated synthetically to D-*erythro*-sphingosine (21).

D-*erythro*-sphingosine   (21)

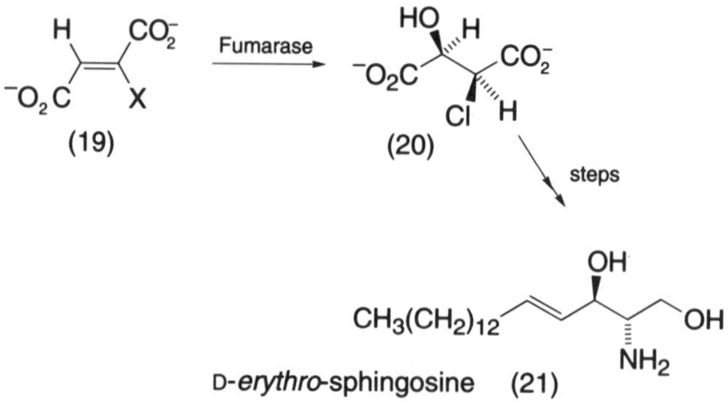

Scheme 5.19.

Fumarase has also been employed in the synthesis of isotopically labelled materials in optically pure form (Scheme 5.20). Thus the fumarase-catalysed addition of D_2O to ^{13}C-labelled fumaric acid yielded the corresponding isotopically labelled malic acid, which proved to be a valuable intermediate for the synthesis of (R)-$[1$-$^{13}C,2$-$^2H]$-malonic acid (22).

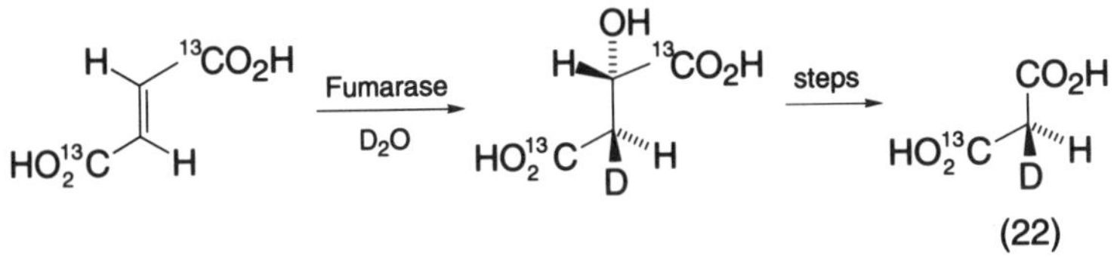

Scheme 5.20.

3-Methylaspartase from *Clostridium tetanomorphum* catalyses the stereospecific addition of ammonia to mesaconic acid (23) to yield $(2S,3S)$-3-methylaspartic acid (24). The latter can then be converted to the isosteric L-valine analogue $(2R,3R)$-3-bromobutyrine (25) (Scheme 5.21).

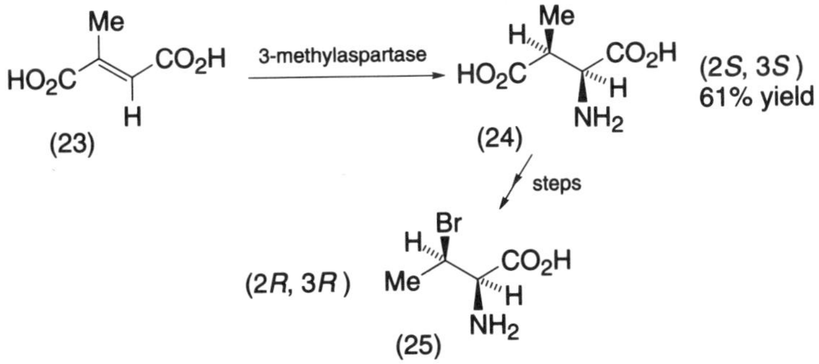

Scheme 5.21.

An unusual enzyme, organomercury lyase, is able to convert carbon–mercury bonds to carbon–hydrogen bonds, as illustrated by the transformation shown in Scheme 5.22.

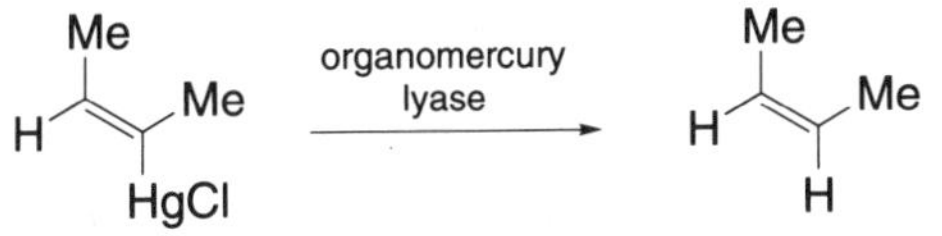

Scheme 5.22.

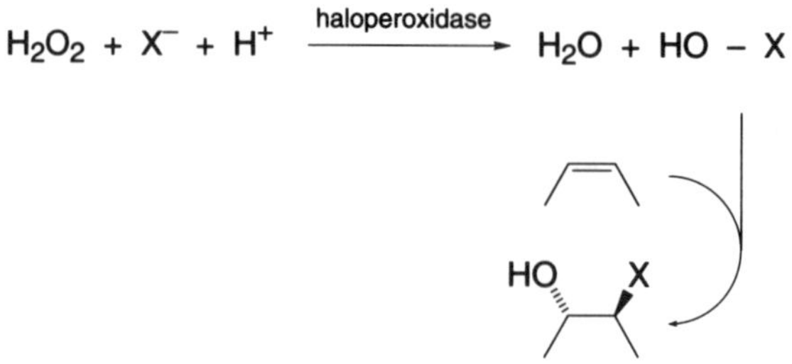

$$H_2O_2 + X^- + H^+ \xrightarrow{\text{haloperoxidase}} H_2O + HO-X$$

Figure 5.1.

Haloperoxidases are potentially very useful catalysts owing to their ability to add HO-X (X = I, Br, Cl) to double bonds to give the corresponding halohydrin. Note that these reactions involve the formation of HO-X, from hydrogen peroxide (H_2O_2) and the halide ion (Figure 5.1), which then adds to the double bond of an alkene. For example, the chloroperoxidase from *Caldariomyces fumago* and the bromoperoxidase from *Corallina pilulifera* have been used to prepare various halogenated nucleic acid bases. Thus uracil can be converted to 5-chloro-, 5-bromo-, or 5-iodouracil (26) by varying the halogen donor (i.e. HCl, HBr, or HI, respectively) in the presence of the chloroperoxidase (Scheme 5.23).

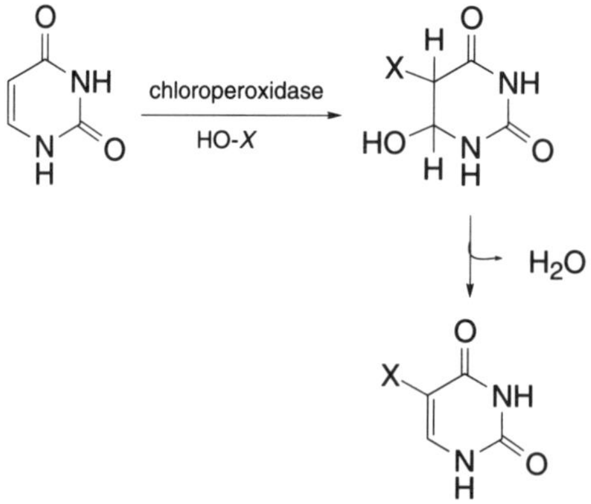

Scheme 5.23.

Finally, in this section on stereospecific construction of C—X bonds, mention must be made of isopenicillin N synthase. This enzyme catalyses the key step in the enzymic synthesis of penicillins [i.e. conversion of δ-(L-α-aminoadipyl)-L-cysteinyl-D-valine (27) to isopenicillin N (28), the precursor to the penicillin antibiotics] (see Chapter 6.5). In this reaction, both a C—N bond and a C—S bond are constructed, both with retention of stereochemistry (Scheme 5.24). This enzyme has a very broad substrate

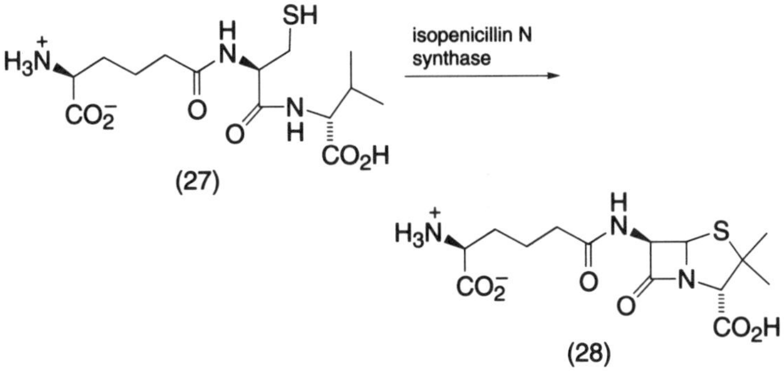

Scheme 5.24.

specificity, particularly with respect to variants in the valinyl moiety of the tripeptide, and it can used to prepare a number of novel penicillins [e.g. (29)] that would be very difficult to obtain by total synthesis.

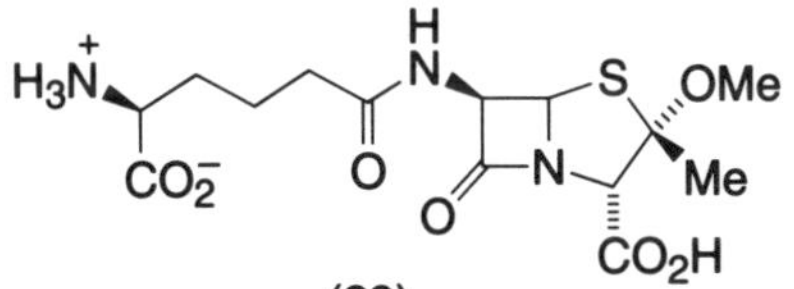

5.4 Application of enzymes in the synthesis of carbohydrate derivatives, including oligosaccharides

The application of enzymes to the selective manipulation of carbohydrates is an area where much work is currently being carried out. The advantage of enzymes here is not simply the opportunity to prepare optically pure chiral intermediates, but perhaps more importantly to carry out highly *regioselective* reactions and thereby overcome the need to employ the complex protection/deprotection strategies that are currently used. This is particularly true in the field of oligosaccharide synthesis, as will be seen later in this section. One carbohydrate-based biotransformation that is an industrially important process is the conversion of glucose (30) to fructose (31) using fructose-glucose isomerase. This seemingly facile reaction is important in the manufacture of high-fructose cornsteep liquor (Scheme 5.25) (see also Section 6.6).

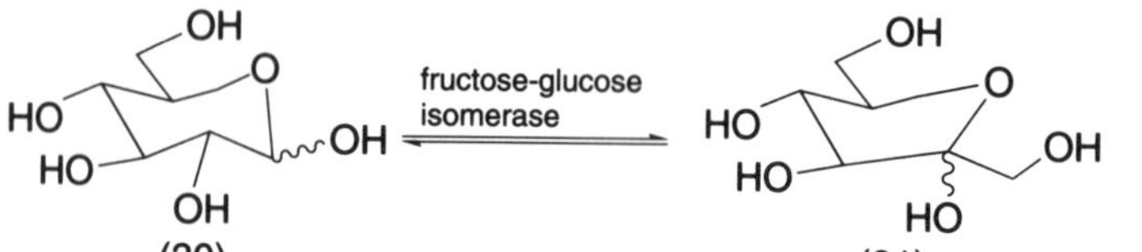

Scheme 5.25.

Many biologically important carbohydrates are phosphorylated, and it is obviously important that the phosphorylation be carried out in a highly regioselective manner. Hexokinase catalyses the phosphorylation of a range of sugars to give the C-6-phosphate, as illustrated for conversion of glucose (30) to glucose-6-phosphate (32) (Scheme 5.26).

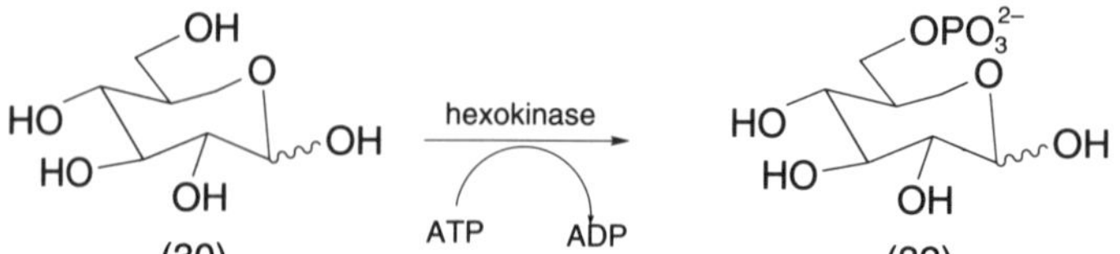

Scheme 5.26.

The enzyme hexokinase, in common with other kinases that catalyse phosphorylation reactions, requires adenosine triphosphate (ATP) as the substrate. The latter is converted to adenosine diphosphate (ADP) during the reaction and must be recycled to avoid the consumption of stoichiometric amounts of the ATP. This can be readily achieved by the introduction of a second enzyme reaction which converts ADP back to ATP. Thus, pyruvate kinase phosphorylates ADP using phosphoenol pyruvate as the phosphate donor, yielding ATP and pyruvic acid. In this way, these phosphorylation reactions can be carried out using a catalytic amount of ATP provided that a stoichiometric quantity of phosphoenol pyruvate is used (Scheme 5.27).

Glycerol kinase uses a simple carbohydrate, glycerol, as its substrate converting it to optically pure *sn*-glycerol-3-phosphate ($R^1 = OH$) (see Chapter 6). The enzyme shows a broad substrate tolerance and has been used to phosphorylate a range of glycerol analogues (33) as shown in Scheme 5.28.

In Chapter 3, it was shown that hydrolase enzymes can be used to catalyse the esterificaton of alcohols using irreversible acylating agents in low-water environments. This methodology has proved to be useful in

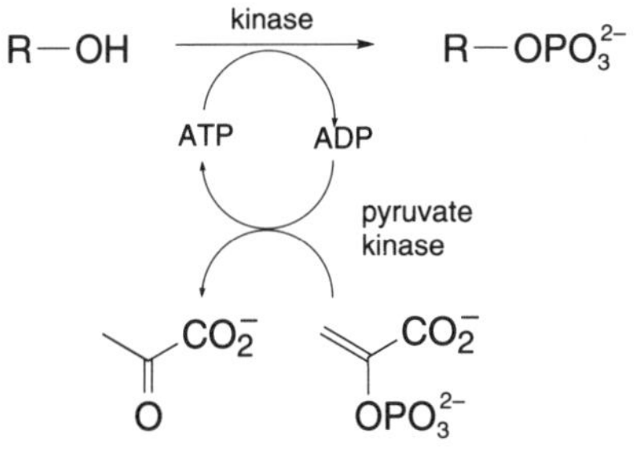

Scheme 5.27.

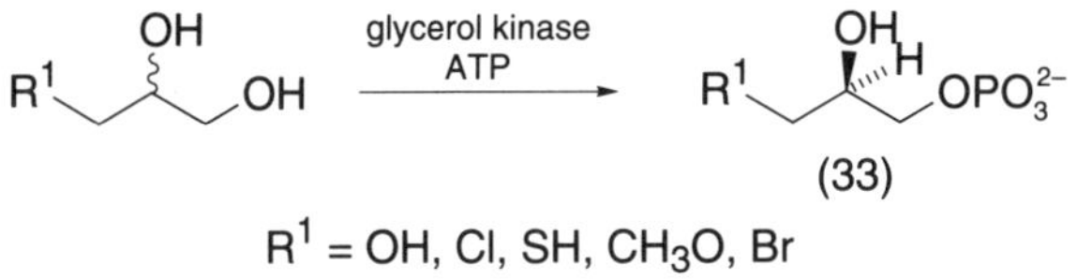

(33)

$$R^1 = OH, Cl, SH, CH_3O, Br$$

Scheme 5.28.

carbohydrate chemistry, particularly in distinguishing hydroxyl groups of similar reactivities, as shown in Scheme 5.29. Thus, using *Pseudomonas fluorescens* lipase in vinyl acetate (acetyl donor), the α-methyl glucoside (34) was selectively acetylated at C-2 (94% yield), with no reaction at C-3, whereas by simply switching to the β-methyl glucoside (35) the ratio of acetylation at C-2/C-3 changed to 7:86. The reason for this control by the anomeric centre on the regioselectivity of acetylation is unclear, but overall this represents a useful synthetic methodology.

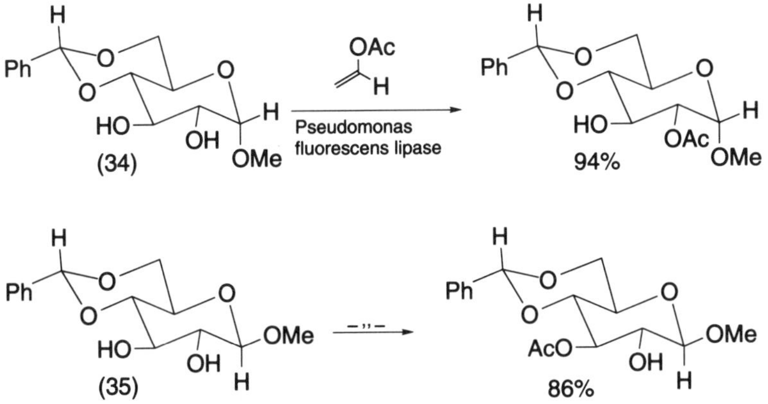

Scheme 5.29.

Enzymes are proving to be extremely useful in the synthesis of glyco-sidic bonds, with exquisite control of both stereoselectivity and regio-selectivity. The challenge in this area is to simultaneously control both the stereoselectivity and regioselectivity of glycoside bond formation. The problem of regioselectivity is clearly illustrated by the fact that if one takes three different amino acids, they can be combined to form six different tripeptides, whereas three different monosaccharides can be combined to give 1,058 different trisaccharides. For four monosaccharides the number rises to over 51,000! Many enzymes, particularly those involved in carbohydrate chemistry, are inherently regioselective and hence look to be attractive catalysts for glycoside synthesis. Essentially two approaches are possible using enzymes, either (i) by using glycosyl transferases or (ii) by exploiting the reverse action of glycosidases. Glycosyl transferases are the family of enzymes that Nature uses for the assembly of oligosaccharides. They require the donor sugar to be present in an activated form as the corresponding nucleotide sugar. This can be illustrated by the synthesis of an artificial sweetener (36) using sucrose synthetase (Scheme 5.30).

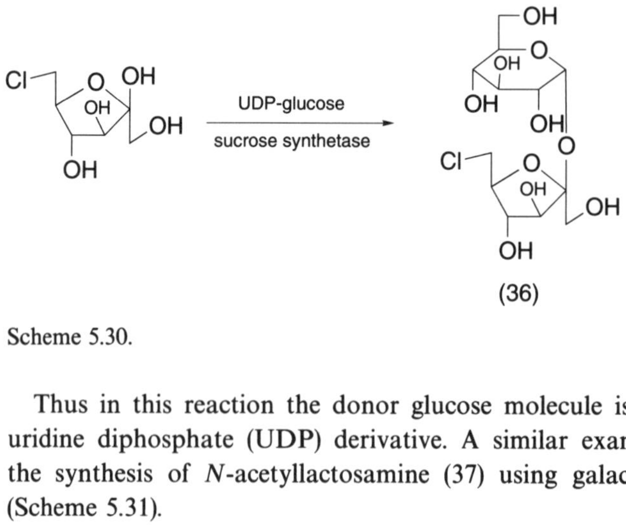

Scheme 5.30.

Thus in this reaction the donor glucose molecule is activated as its uridine diphosphate (UDP) derivative. A similar example is given by the synthesis of *N*-acetyllactosamine (37) using galactosyl transferase (Scheme 5.31).

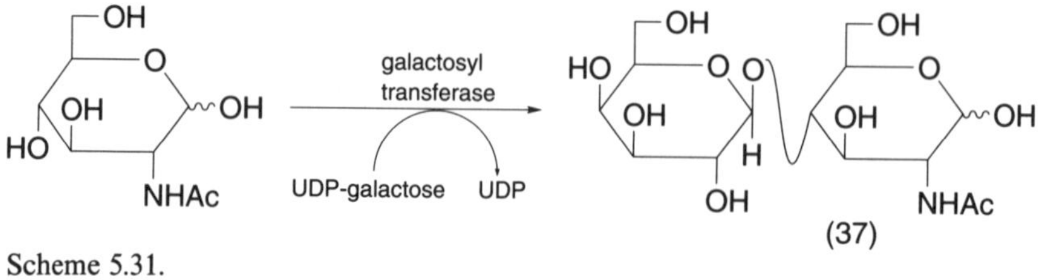

Scheme 5.31.

In these glycosyl transferase reactions, a consequence of sugar transfer is that the nucleotide diphosphate (XDP) is released into solution. However, by the addition of further enzymes it is possible to convert the XDP back into the nucleotide-activated sugar ready for the glycosyl transferase reaction (Scheme 5.32). Thus, in this multienzyme reaction, a total of six enzymes is used to convert glucose-6-phosphate (32), which is cheap and readily available, into N-acetyllactosamine (37), via the in situ generation of UDP-glucose and UDP-galactose.

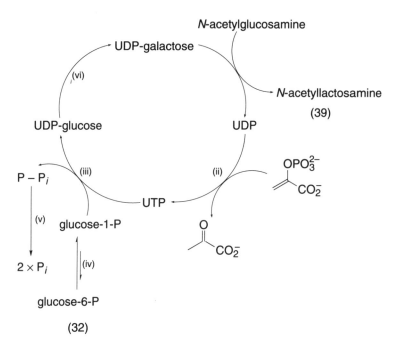

enzymes: (i) galactosyl transferase, (ii) pyruvate kinase,
(iii) UDP-glucose pyrophosphorylase, (iv) phosphoglucomutase,
(v) inorganic pyrophosphatase, (vi) UDP-galactose epimerase

Scheme 5.32.

A similar approach has been used to prepare oligosaccharides that contain sialic acid (N-acetylneuraminic acid) (8). We saw earlier that sialic acid itself can be prepared using an aldolase reaction, and this forms the starting point for another cascade of enzyme-catalysed reactions. In this instance, sialic acid is firstly activated as its cytidine monophosphate (CMP) derivative via coupling with cytidine triphosphate (CTP) in the

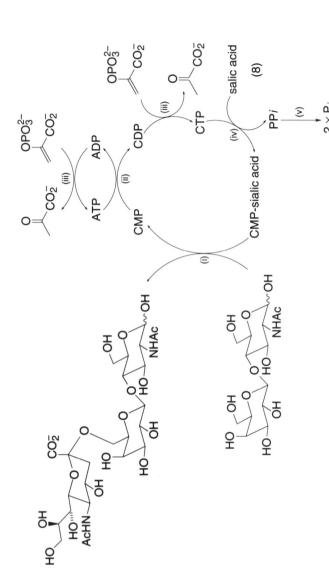

enzymes: (i) α-2,6-sialyltransferase, (ii) nucleoside monophosphate kinase or myokinase, (iii) pyruvate kinase, (iv) CMP-sialic acid synthetase, (v) inorganic pyrophosphatase

Scheme 5.33.

presence of CMP–sialic acid synthetase. The CMP–sialic acid thus generated is then a substrate for a sialyl transferase that incorporates sialic acid into the growing oligosaccharide chain, releasing CMP, which can be recycled back to CMP–sialic acid in a fashion analogous to that described earlier (Scheme 5.33).

The alternative method for stereoselective glycoside synthesis exploits the "reverse" activity of glycosidases. Glycosidases are enzymes that under normal conditions catalyse the *hydrolysis* of oligosaccharides via cleavage of the glycosidic linkage. An example is given by the hydrolysis of lactose (38) to give galactose (39) and glucose (30) (Scheme 5.34). The formal

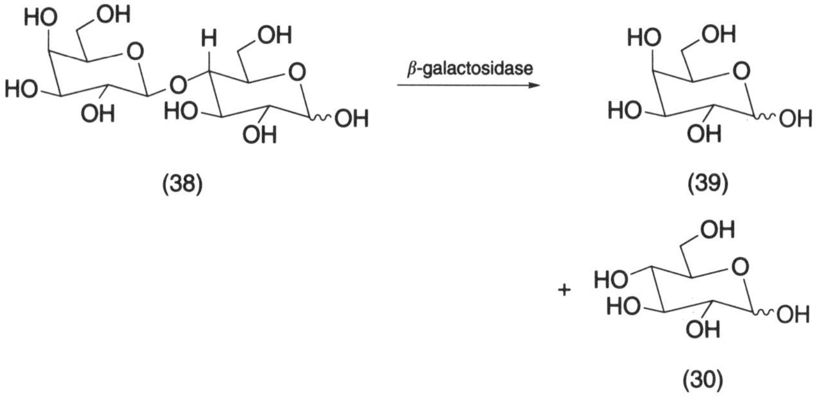

Scheme 5.34.

reversal of this reaction (i.e. the combination of galactose and glucose to yield lactose) is unsuccessful owing to the unfavourable equilibrium constant of the reaction. However, it is possible to exploit a kinetic approach by using a reactive sugar donor (usually an *o*- or *p*-nitrophenyl glycoside) in place of the simple sugar. In this way, good yields of the disaccharide lactose (38) can be obtained, particularly if the acceptor sugar is present in high concentrations (Scheme 5.35).

These reactions can be rationalized by envisaging a glycosyl cation intermediate that is subject to competitive interception either by the hydroxyl group of another sugar, leading to disaccharide formation, or by water, leading to overall hydrolysis (Figure 5.2).

This basic reaction has been exploited in a variety of ways, for the synthesis of both simple glycosides (Scheme 5.36) and more complex oligosaccharides (Scheme 5.37). In Scheme 5.37, *p*-nitrophenyl-α-L-fucose

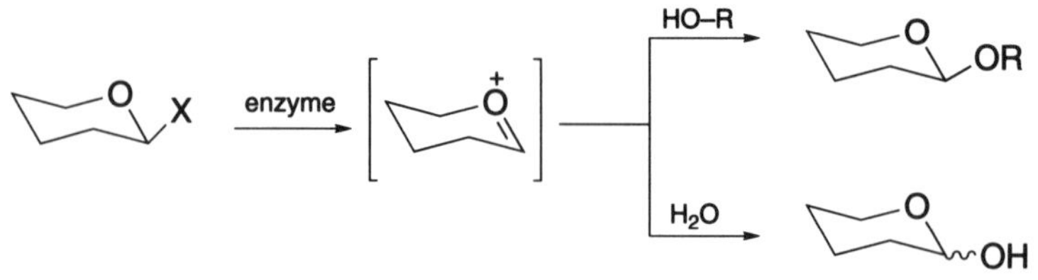

Figure 5.2.

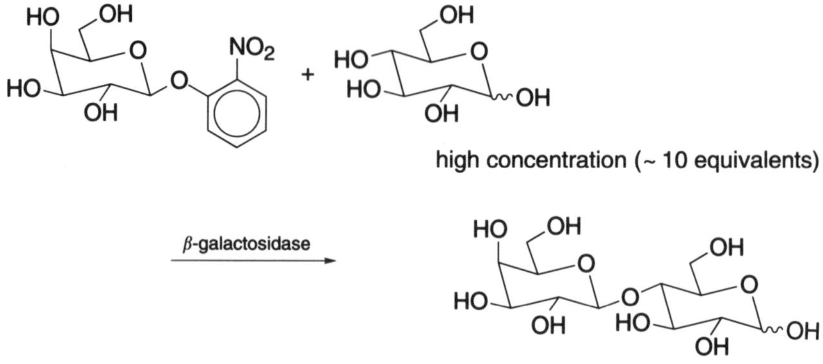

high concentration (~ 10 equivalents)

(38) 40–50% yields

Scheme 5.35.

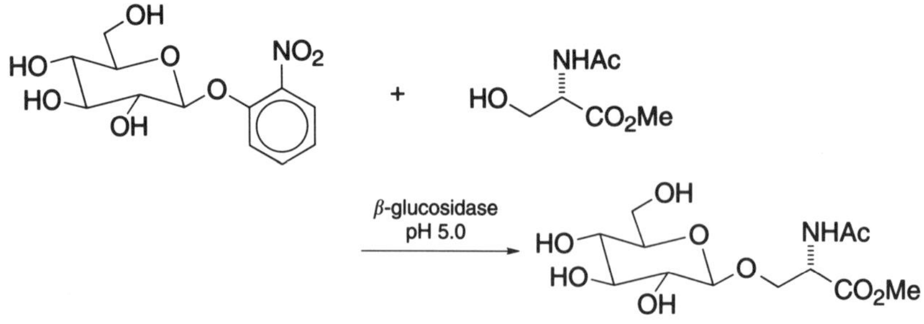

Scheme 5.36.

(40) is used as the glycol donor, leading to two products that differ in the regioselectivity of glycosidic bond formation.

5.5 Conclusions and overview

This chapter has surveyed a wide range of C—X bond-forming reactions, where X = C, O, N, S, and halogen. The enzymes that catalyse carbon–carbon bond formation are potentially the most useful and show some

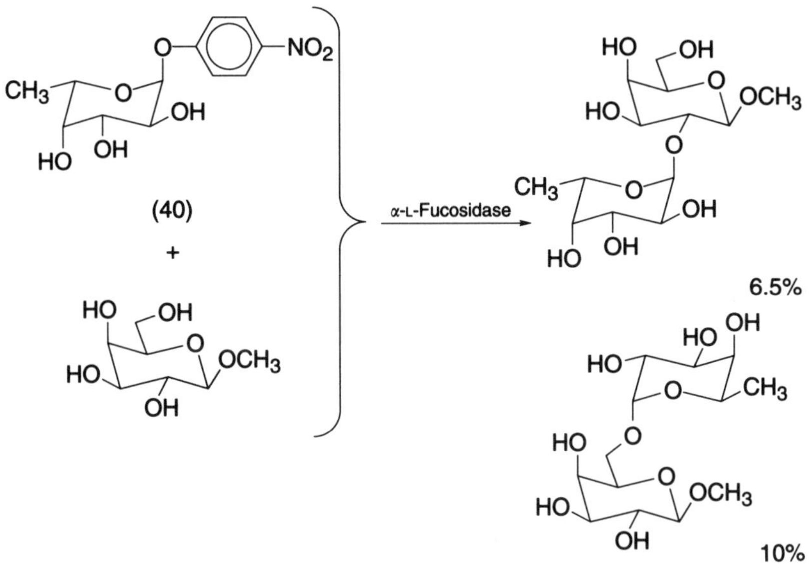

Scheme 5.37.

versatility in the type of substrate they can use. Undoubtedly the advances being made in the overexpression of genes in user-friendly host organisms, such as *Escherichia coli*, will make aldolases much easier to obtain in large quantities. The application of enzymes for the synthesis of carbohydrate-type materials is certain to expand. In tandem with the discovery of new therapeutic targets, wherein carbohydrates play a fundamental recognition rôle, the development of highly selective and efficient methods for synthesis will be required. The high levels of regioselectivity and stereoselectivity displayed by enzymes such as glycosyl transferases and glycosidases make them attractive catalysts for selective synthesis and may offer advantages over existing methods in organic chemistry.

Bibliography

Bednarski, M. D., & Simon, E. S. (1991). *Enzymes in Carbohydrate Synthesis.* American Chemical Society, Washington, D.C.

Drueckhammer, D. G., Hennen, W. J., Pederson, R. L., Barbas, C. F., III, Gautheron, C. M., Krach, T., & Wong, C.-H. (1991) Enzyme Catalysis in Synthetic Carbohydrate Chemistry. *Synthesis*, 499.

Neidleman, S. L., & Geigert, J. (1986). *Biohalogenation – Principles, Basic Rôles, and Applications.* Ellis Horwood, Chichester.

Toone, E. J., Simon, E. S., Bednarski, M. D., & Whitesides, G. M. (1989). Enzyme-catalysed Synthesis of Carbohydrates. *Tetrahedron, 45*, 5365.

Wong, C.-H. (1989). Enzymatic Catalysts in Organic Synthesis. *Science, 244*, 1145.

6

The application of biocatalysis to the manufacture of fine chemicals

6.1 Introduction

Biological catalysis has such a pervasive influence on the industries which are closely associated with our daily lives that it is difficult to contain a short review within sensible bounds. The traditional large-scale crafts associated with agriculture, the manufacture of foods and drinks, and the preparation and cleaning of fabrics all use biological catalysis. At the other extreme, biological catalysis has an important rôle to play in analytical chemistry. Enzymes are used to monitor the levels of metabolites as an important feature in the diagnosis and therapy of disease. These techniques have been extended to the use of enzymes to amplify the responses in antibody-based assays, as well as to mediate the detection of an environmental pollutant at an electrode. The same catalysts also play a rôle in the degradation of pollutants, which is, perhaps, simply an extension of the rôle of biochemistry applied to the traditional craft of cleaning.

However, it is the use of enzymes in the large-scale manufacture of synthetic organic chemicals which is the main topic of this chapter. The recent interest amongst organic chemists arises from the high degrees of specificity and selectivity that characterize enzyme catalysis. That this is a valuable resource in manufacture cannot be doubted, as judged by the value of the products which it produces. Nevertheless, the methodology should not be oversold. Biocatalysis has a part to play in the modern chemical industry, but it also has limits which it is necessary to understand.

6.2 Fermentation

For some years the syntheses of ascorbic acid (vitamin C) (1) and D-ephedrine (2) (see Chapter 1) remained isolated examples of the controlled use of biological catalysis outside of processes which make use of a "wild-type"

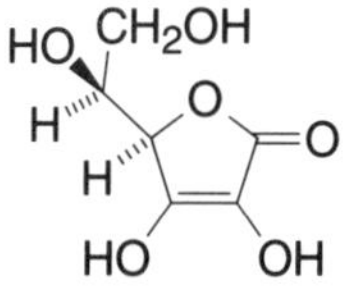

CH3
H——NHCH3
H——OH

HO CH2OH
H''' O
H'''   =O   (1)
HO   OH   (2)

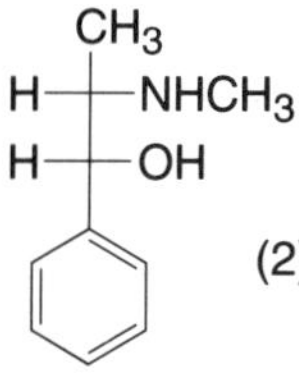

ascorbic acid   D-ephedrine

fermentation. Yet any review of the industrial use of biocatalysis ought to note the latter fermentations, however briefly. Not only are they important industrial processes for the production of commodity chemicals, but there is also little in concept which distinguishes the fermentative conversion of ethanol to acetic acid in the manufacture of vinegar, or the manufacture of ethanol from glucose, from the conversion of D-glucitol (3) to L-sorbose (4) in ascorbic acid manufacture (Scheme 6.1). In the wild-type fermentative

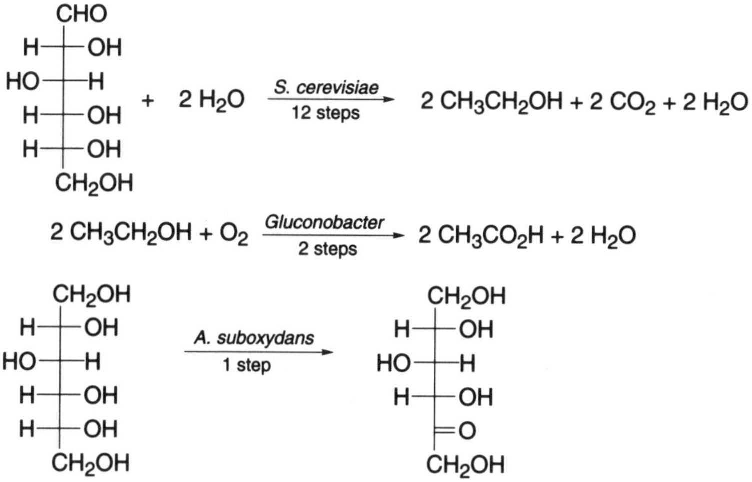

CHO
H——OH
HO——H
H——OH   + 2 H2O $\xrightarrow[\text{12 steps}]{\textit{S. cerevisiae}}$ 2 CH3CH2OH + 2 CO2 + 2 H2O
H——OH
CH2OH

2 CH3CH2OH + O2 $\xrightarrow[\text{2 steps}]{\textit{Gluconobacter}}$ 2 CH3CO2H + 2 H2O

CH2OH   CH2OH
H——OH   H——OH
HO——H $\xrightarrow[\text{1 step}]{\textit{A. suboxydans}}$ HO——H
H——OH   H——OH
H——OH   =O
CH2OH   CH2OH

(3)   (4)

Scheme 6.1.

conversions with *Gluconobacter* sp. to produce acetic acid, or with *Saccharomyces cerevisiae* to produce ethanol, the organisms are allowed to grow on the substrates before switching to the conversions proper. Similarly, in the latter instance *Acetobacter suboxydans* is allowed to grow on the D-glucitol before the process switches to its oxidation, in which L-sorbose is the product. The essential difference is that there are several catalytic steps between substrate and product in the former examples, whereas there is only one in the latter.

6.2.1 Organic solvents

Despite the efficiency of the manufacture of ethanol from petrochemical feedstocks, much of the world's production is based on a fermentation process. Over the past 75 years in the United States, where the total annual production now stands at just under 4 million tonnes, the source of this basic chemical feedstock has swung away from fermentation to petro- chemistry and back again (Table 6.1). The carbon source for the fermenta- tion is glucose derived from starch (see Section 6.6). An even larger quantity, about 9.5 million tonnes, is produced each year in Brazil from cane sugar. Nowadays the prime consumer is the motor car.

Economic, political and social pressures have influenced the manufacture of ethanol since the fermentation process was introduced many years ago (see Chapter1). In the nineteenth century its status as an industrial solvent was assured when industrial methylated spirit, which is ethanol denatured with methanol to make it unfit for human consumption, was freed from excise duty. In the twentieth century, the social and political pressures which provide economic support for agriculture have allowed fermentation to compete effectively with petrochemistry as a manufacturing process. The technical achievements in chemistry, biology and engineering have only facilitated the switch between carbohydrate and oil as feedstock for the process.

The manufacture of acetone shows a different interplay. Its source in the nineteenth century was wood, whose destructive distillation yielded both acetone and methanol (Scheme 6.2). In the United Kingdom the necessary timber was imported from Austria, and that supply was cut off during World War I. Weizmann then developed a fermentation process based on the growth of *Clostridium acetobutilicum* (then called *Bacillus macerans*) on starch in the absence of oxygen. This yielded a mixture of acetone, butan-1-ol and ethanol in the ratio 6:3:1 by weight; these three

Table 6.1. *Proportion of ethanol which is manufactured by fermentation in the United States*

Year of manufacture	1920	1935	1954	1963	1977	1982	1991
% from fermentation	100	90	30	9	6.5	55	94

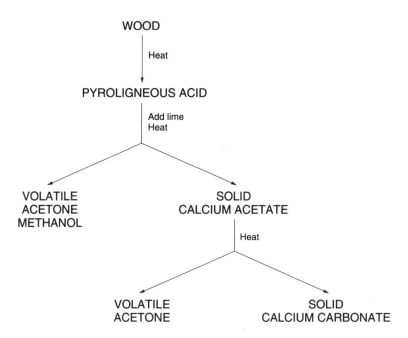

Scheme 6.2.

solvents have boiling points which are sufficiently different to allow them to be separated by distillation. Subsequent development led to a set of processes, with different strains of clostridia yielding acetone, butan-1-ol and propan-2-ol as separate products. The processes do make less efficient use of the input carbohydrate than does ethanol fermentation; the output of solvent is about 60% of the theoretical maximum, compared to about 80% for ethanol.

The microbiological method soon found itself in competition with the growing petrochemical industry, and in the United Kingdom the use of fermentation for acetone manufacture was stopped in 1957. However, the method is still used in some parts of the world, notably in China, where oil is in short supply. Given the present dominance of fermentation in the ethanol market in the United States, it would not be surprising to see some of their current needs for acetone, butan-1-ol and propan-2-ol production (Table 6.2) being supplied by a return to fermentation.

6.2.2 Carboxylic acids

Fermentation processes are the sources of a number of carboxylic acids, of which the largest bulk products are citric (5) and the amino acids

Table 6.2. *Production of aliphatic alcohols and acetone in the United States in 1991*

Route	Methanol	Ethanol	Propan-2-ol	Butan-1-ol	Acetone
Chemical	3.9[a]	0.25	0.59	0.60	0.97
Fermentation	—	3.6	—	—	—

[a]Million tonnes per year.
Source: Chem. Eng. News (1992), June 29, p. 36; November 2, p. 7.

(5) citric acid   (6) L-glutamate   (7) L-lysine

L-glutamate (6) and L-lysine (7). The annual world production of citric acid is now well over 0.5 million tonnes. Indeed, this figure probably underestimates the output in the Far East, particularly in countries such as China. Some estimates place the world production as high as 0.9 million tonnes, representing a sustained annual growth of over 8.5% for the past 60 years (Table 6.3).

Although citric acid is a relatively simple achiral molecule of only six carbon atoms, for its manufacture there is no effective synthetic chemistry which can compete, on a large scale, with fermentation. Even the scale of the fermentation itself is large, with vessels of $500 \, m^3$ now in use. The wide range of uses for citric acid, in processes as diverse as an acidity regulator in foods, as a cleaning agent for metal surfaces, as a replacement for phosphate in detergents, or as an agent to control the viscosity of the muds used in drilling for oil, is possible because the fermentation process is so effective.

Several amino acids are manufactured in large quantities (Table 6.4). They are mostly used as food ingredients, for example to supplement

Table 6.3. *Annual production of citric acid*

Year	1929	1943	1950	1976	1983	1991
Annual production (tonnes)	5×10^3	12×10^3	50×10^3	0.2×10^6	0.37×10^6	0.55×10^6

Table 6.4. *Manufacture of natural amino acids with an annual production of over 500 tonnes*

Amino acid	Method of manufacture	Annual tonnage (1987)
L-glutamic acid	Microbiological	340,000
DL-methionine	Chemical	250,000
L-lysine	Microbiological	70,000
Glycine	Chemical	6,000
L-aspartic acid	Enzymatic	4,000
L-phenylalanine	Microbiological	3,000
DL-alanine	Chemical	1,500
L-cysteine	Extraction or enzymatic	1,000
L-arginine	Microbiological	1,000
L-glutamine	Microbiological	850

animal feed. Since the D-enantiomers are inverted in the animal to the metabolically active L-enantiomers, the chemically manufactured racemate often can replace the natural isomer derived from a fermentation process. The choice of route can be made largely on economic grounds. In contrast, the supply of starting materials for other compounds, such as the low-calorie sweetener Aspartame (8) or the β-lactam antibiotic imipenem (thienamycin) (9), demands a chiral input. Chemical synthesis is employed

(8)   (9)
Aspartame   imipenem

to produce DL-methionine (10) (Scheme 6.3) and glycine, but many of the other valuable amino acids are the products of fermentation or of processes dependent on the use of enzymes (see Section 6.4).

6.2.3 Antibiotics

Following the discovery of penicillin, the antibiotics industry has had a major influence on the development of fermentation processes. Well over

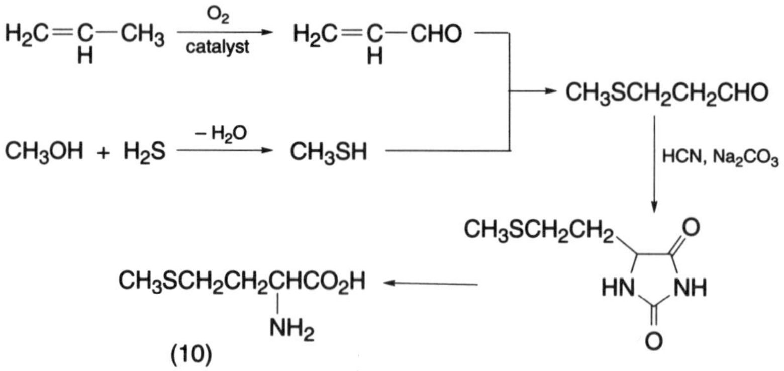

Scheme 6.3.

one hundred compounds have been fermented on a commercial basis, mostly for use in human and veterinary medicine and in agriculture, but the requirement for none of these materials is as large as for citric acid, or for the amino acids. Many are prepared as crude bulk products, particularly those such as kanamycin (11) or blasticidin (12), which are used in

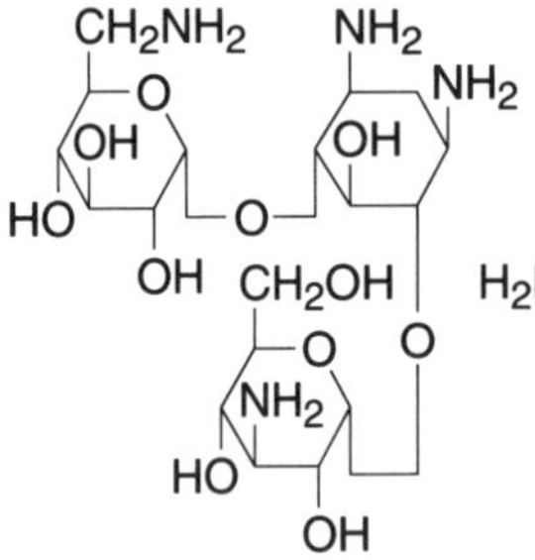

kanamycin (11) 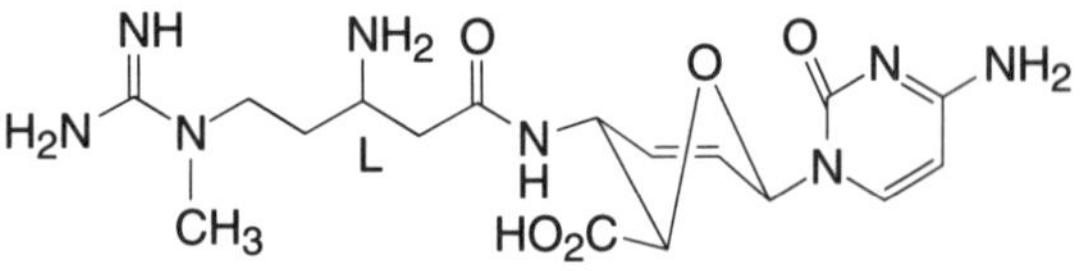   blasticidin S (12)

agriculture. Generally their value as chemical entities, particularly for the pharmaceutical industry, does not lie in the fermentation products themselves, but in the compounds which are made from them. Those made from the 25,000 tonnes of penicillin produced each year are particularly valuable (Table 6.5). In that context, their value as fermentation products is rivalled only by the recombinant proteins (see Section 6.11).

What this survey of fermentation products should illustrate is (a) the scale of production, which can rival that of the bulk chemical industry, (b) the range of chemical entities, from ethanol to antibiotics, which can be produced, and (c) the value of the industry, particularly in supplying

Table 6.5. *Drugs derived from low-molecular-weight fermentation products which, in 1990, were amongst the world's top-50 best-selling pharmaceuticals*[a]

Source material	Drug name		Sales ($M)	Manufacturer
	Trade	Generic		
Penicillin G or V	Ceclor	Cefaclor	837	Eli Lilly
	Amoxil	Amoxicillin	442	Smith Kline
Penicillin G & clavulanic acid	Augmentin	Amoxicillin & clavulanic acid	710	Beecham Smith Kline Beecham
Cephalosporin C	Fortam	Ceftazidime	373	Glaxo
	Rocephin	Cephtriaxone	665	Roche
	Claforan	Cefotaxime	485	Hoechst
Chemical synthesis	Primaxin	Imipenem & cilastatin	398	Merck
Erythromycin	Erythrocin	Erythromycin	421	Abbott
Lovastatin	Mevacor	Lovastatin	751	Merck
Cyclosporin	Sandimmum	Cyclosporin	420	Sandoz

[a] All are antibiotics except for lovastatin and cyclosporin. Imipenem (thienamycin) is also a fermentation product, but it is more conveniently manufactured by chemical synthesis.
Source: Scrip.

starting materials for chemical synthesis. It is also noteworthy that for the β-lactam antibiotics the programmes of chemical synthesis are dependent on the use of enzymes at a number of other key stages (see Sections 6.4.2 and 6.5).

6.3 Steroids

Steroids are animal hormones. However, their potential as pharmaceutical products could not have been realized while their manufacture depended on methods such as the extraction of estrone (13) from the urine of pregnant mares! Cholesterol (14) and bile acids were better sources, but one of the initial steps in the requisite conversion involved oxidation with chromic acid, a transformation which was not efficient. Not until Marker realized in the early 1940s that progesterone (15) and testosterone (16) could be readily synthesized from diosgenin (17), a plant steroid obtained from a

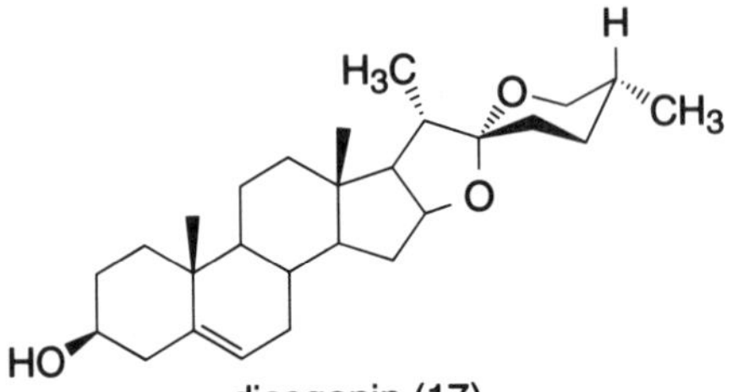

diosgenin (17)

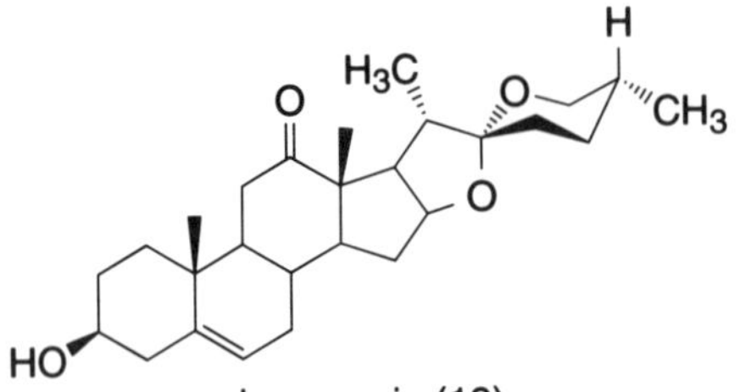

hecogenin (19)

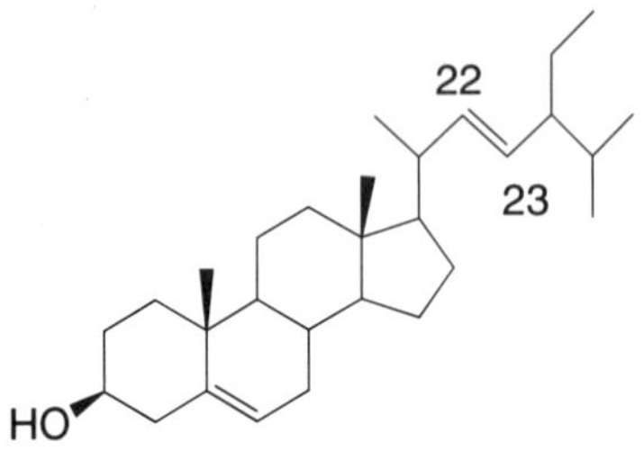

stigmasterol (23)

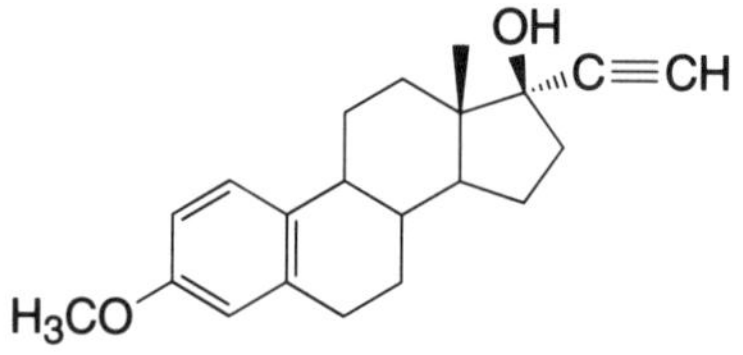

(β-sitosterol is 22,23-dihydrostigmasterol) (24)

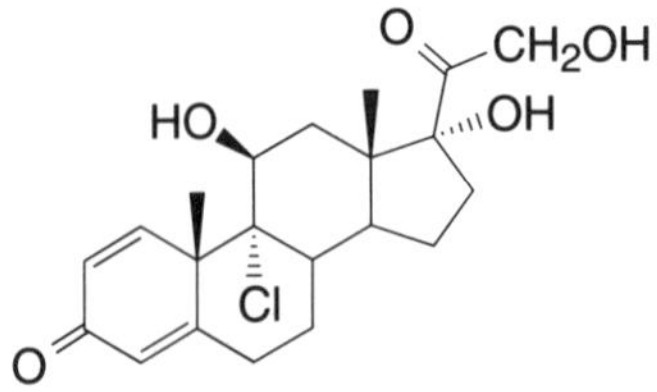

mestranol (22)

beclamethasone (20)

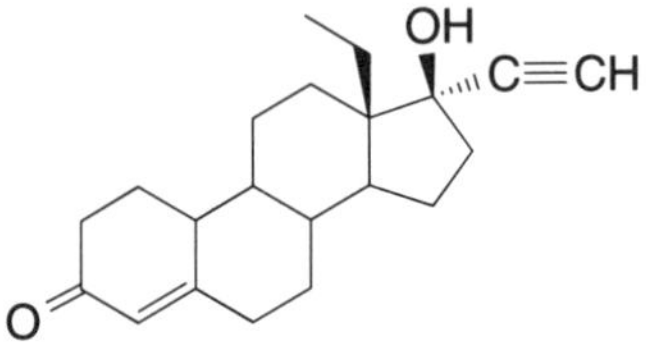

norgestrol (26)

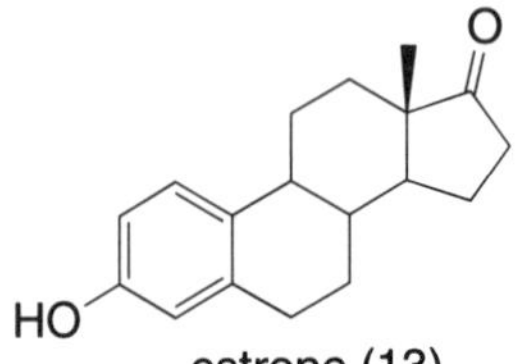

estrone (13)

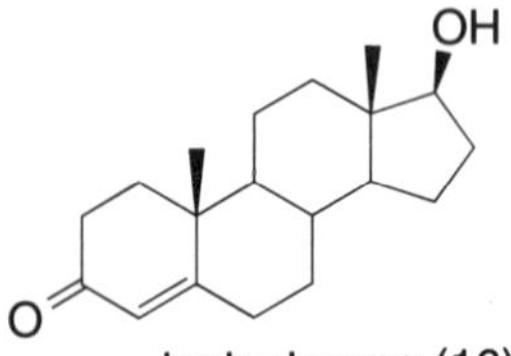

testosterone (16)

progesterone (15)

cortisone (18)

androstenedione (25)

hydrocortisone (21)

cholesterol (14)

member of the yam family (*Dioscorea mexicana*) which grows wild in Mexico, did the programmes of synthetic chemistry develop.

Diosgenin does not contain an oxygen atom attached to C-11 or C-12 of the steroid nucleus, and this proved to be a major problem in the synthesis of the corticosteroids such as cortisone (18). A search for other suitable starting materials yielded hecogenin (19), which is present in the African sisal plant (*Agave sisalana*). This natural product contains an oxygen atom at C-12, and the functionality can be transferred to the adjacent carbon atom, C-11. However, about six chemical steps are required to move the oxygen atom from C-12 to the β-orientation at C-11, a substitution pattern which is a common feature in the anti-inflammatory agents, such as beclamethasone (20).

The process was simplified in 1952 when Petersen and Murray described a direct hydroxylation at C-11 catalysed by the fungus *Rhizopus arrhizus*. Although this oxidation produced the 11α-hydroxysteroid, inversion of the configuration of the hydroxyl group is relatively straightforward. The process can reasonably be credited with reviving interest in the application of biological catalysis, and it had a significant effect on the future of steroid manufacture.

Although the 11α-hydroxylation reaction became a commercial process, it is not an efficient one. It has a high selectivity, but a very low catalytic intensity. The fungus is first grown as if in a normal fermentation. The steroid, which is very poorly soluble in the aqueous fermentation medium, is added as a fine dispersion to a final concentration between 0.1% and 1% (w/v). The catalyst itself is a cytochrome P-450 mono-oxygenase attached to an intracellular membrane. It is an enzyme complex which also requires the co-factor NADPH to reduce one atom of molecular oxygen to water, while the other oxygen atom is incorporated into the steroid (Scheme 6.4).

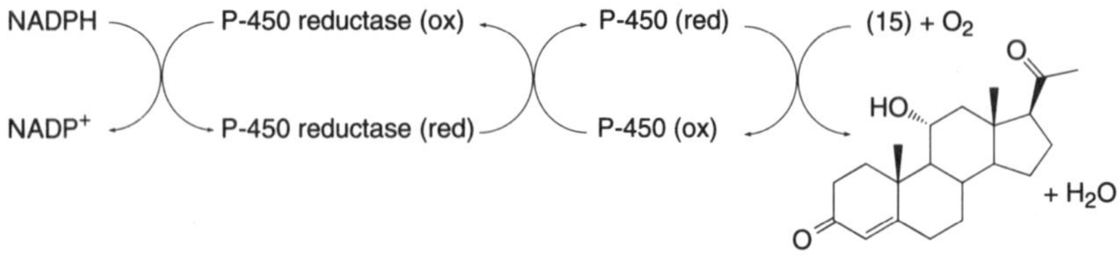

Scheme 6.4.

The dissolution of the solid substrate, its transport into the cell, and the transport of the product back out slow the catalysis, and the process takes several days to complete. The $NADP^+$ is reduced back to NADPH inside the cell. Other positions on the steroid are hydroxylated, notably the 6α position, if the steroid is left too long in contact with the cells.

Treatment of selected steroids with the micro-organism *Curvularia lunata* will introduce the 11β-hydroxyl group directly, and this is the basis of a commercial process catalysed by a micro-organism which converts 17α-acetoxy-11-deoxycortisol to cortisol (hydrocortisone) (21). The reaction needs more careful control than the 11α-hydroxylation because other carbon atoms in the steroid nucleus are also oxidized, notably to give the 9α-OH and 14α-OH derivatives. Micro-organisms are now known which will hydroxylate almost any available position on the steroid nucleus, but the only other important hydroxylation process in manufacture is that inducing oxidation at C-16, producing, for example, 9α-fluoro-16α-hydroxycortisol in a reaction catalysed by *Streptomyces roseochromogenes*.

$Δ_1$-dehydrogenation is catalysed by the organism *Arthrobacter simplex*. This reaction is useful in the production of the aromatic A-ring in the 19-norsteroids such as estrone (13) or mestranol (22), which lack the 19-methyl group. It is also a reaction which is compatible with the presence of an immiscible organic solvent. This provides a second liquid phase in which the steroid is soluble. In this instance the cells of *A. simplex* will tolerate the presence of toluene, allowing larger amounts of steroid to be conveniently added to the reaction.

Two other useful microbial conversions further illustrate how factors outside the control of the chemical industry affect manufacturing processes. During the 1970s, the Mexican government began to control the collection of the wild tubers from which diosgenin was extracted, and by 1976 their price had risen over 20-fold to $5,600 per tonne. That led to a search for other sources of supply. Stigmasterol (23) recovered from soya oil was already used as an alternative, and the 17β side chain of this sterol is readily converted to the 2-carbon unit of progesterone (15). However, stigmasterol occurs naturally as a mixture with the more abundant β-sitosterol (24), which cannot be employed in the same way because the side chain is saturated. As stigmasterol replaced diosgenin in steroid manufacture, the β-sitosterol accumulated as a waste by-product.

Studies of microbial growth in which organisms were forced to use a steroid, such as β-sitosterol, as their sole carbon source revealed that they would often start by oxidizing the side chain right back to the steroid

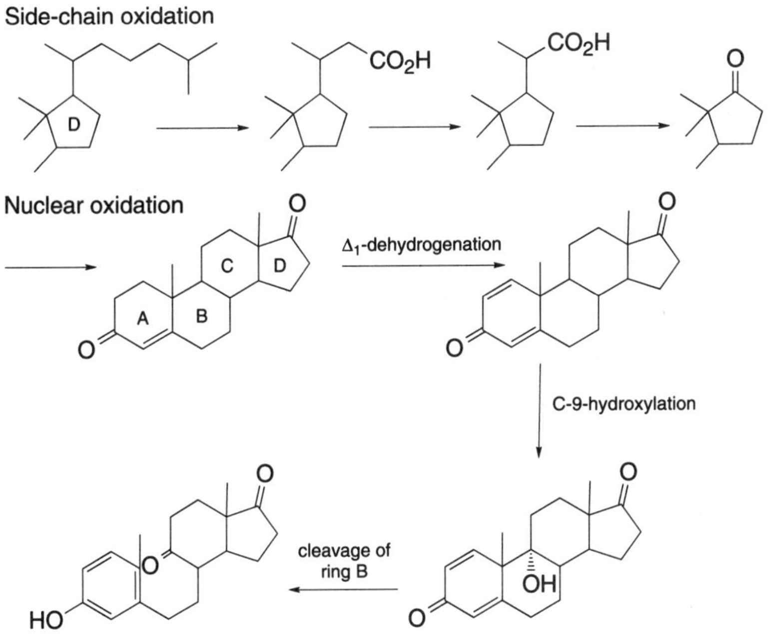

Scheme 6.5.

nucleus, leaving a 17-keto steroid as the product. Further attack was then likely at C-9 (hydroxylation) and at C-1 (Δ_1-dehydrogenation) (Scheme 6.5). That undesirable attack on the nucleus could be inhibited by adding heavy metals to the conversion medium, or by chelating the dissolved iron with α,α'-dipyridyl. Later, some strains of micro-organisms which lacked the two mischievous enzymes were also isolated. The fully developed conversion will oxidize β-sitosterol to androstenedione (25), a useful starting material for synthesis of the contraceptive steroids. As a result, the waste which had accumulated from the use of stigmasterol became a valuable resource, and was known in one company as their "sitosterol mine".

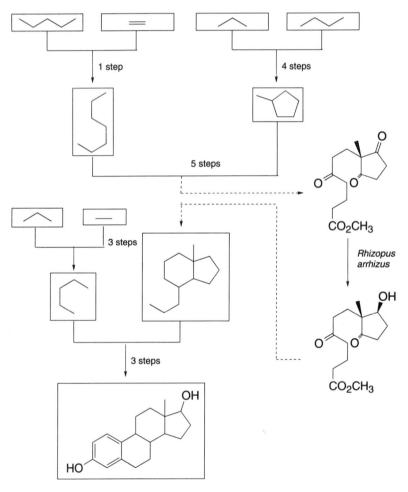

Scheme 6.6.

In at least one other company the increased price of diosgenin stimulated the research to achieve a total chemical synthesis of the steroid nucleus. This strategy has some advantages, one of which is that it makes available compounds based on the 18-methyl-19-norsteroid nucleus, such as nor-gestrol (26), which cannot be prepared from a natural product. As part of the synthetic process, which is a convergent one (Scheme 6.6), the prochiral cyclopenta-1,3-diones can be selectively reduced either by *Rhizopus arrhizus* or by *Saccharomyces uvarum*. There are also chemical strategies which will close the C-ring of the steroid with the correct stereochemistry at C-17.

6.4 Manufacture of amino acids

There are several alternative routes for the synthesis of chiral amino acids. Most are resolutions of a chemically prepared racemate, but there are also direct methods of synthesis from achiral precursors. On the large scale, both methods are represented, although some of the more complex processes probably remain as pilot-plant demonstrations rather than as important manufacturing processes.

6.4.1 L-methionine

The enzymatic method for the resolution of DL-methionine was introduced in the 1950s. Cells of the mould *Aspergillus oryzae* were used to hydrolyse *N*-acyl-DL-methionine. The acylase is specific for the L-isomer, the products being L-methionine and *N*-acyl-D-methionine. After removing the cells, the pH of the filtrate was adjusted to precipitate the L-methionine at its isoelectric point, leaving a solution of the remaining *N*-acyl amino acid. The latter can be racemised chemically and recycled through the process (Scheme 6.7). This basic strategy of hydrolysis and racemisation is typical of simple resolution procedures.

The cells of *A. oryzae* are not a convenient source of the hydrolase enzyme, and their use is hardly an advance on the technology which Pasteur had introduced about a century earlier (see Chapter 1). It is not an active "fermentation" process like the steroid hydroxylations. The cells are grown and harvested before being added to the conversion mixture; consequently, these cells will contain not only the enzyme which is catalysing the hydrolysis but also many others, some of which could oxidize the methionine once it is formed, or catalyse the release of soluble material from the cells. All of these impurities simply contaminate the reaction mixture. The cells recovered from the reaction mixture could be used again in a subsequent

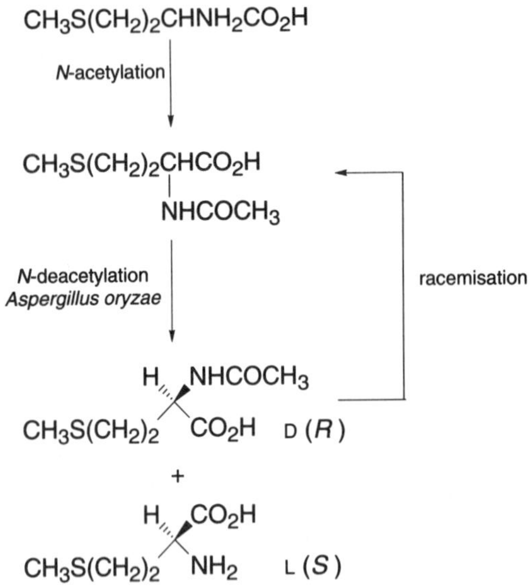

CH$_3$S(CH$_2$)$_2$CHNH$_2$CO$_2$H

    N-acetylation

CH$_3$S(CH$_2$)$_2$CHCO$_2$H
                |
            NHCOCH$_3$

    N-deacetylation   racemisation
    Aspergillus oryzae

            H   NHCOCH$_3$

CH$_3$S(CH$_2$)$_2$ CO$_2$H   D (*R*)

                +

            H   CO$_2$H

CH$_3$S(CH$_2$)$_2$ NH$_2$ L (*S*)

Scheme 6.7.

conversion, but their catalytic activity falls quite rapidly as the cells begin to lyse and decompose. For the hydroxylation of steroids, with its complex enzymology requiring the regeneration of NADPH, these impurities are an acceptable trade-off against the requisite oxidation, and the cells will not be re-used once the oxidation is complete. This is not true of a simple hydrolysis catalysed by one enzyme.

The resolution of methionine probably was the first manufacturing process where these issues were considered and confronted. The enzyme was isolated from the cells and immobilized on a basic ion-exchange resin. The enzyme retains its activity while bound to the resin, which can be handled as a solid-phase catalyst. In this state it is packed into a column through which the reaction mixture is pumped. As the hydrolysis takes place, the pH of the solution falls and needs to be adjusted to prevent the methionine from precipitating. If properly controlled, the resolution can be transformed into a continuous process (Figure 6.1), and one which is much more convenient and hygienic than the original process with separate batches of *A. oryzae* cells. It is not a large-scale process, with only some 150 tonnes of L-methionine being produced each year.

6.4.2 4-hydroxy-D-phenylglycine (27)

Systems similar to that described in Section 6.4.1 are used for the resolution of other racemic amino acid derivatives; the preparation of D-amino acids

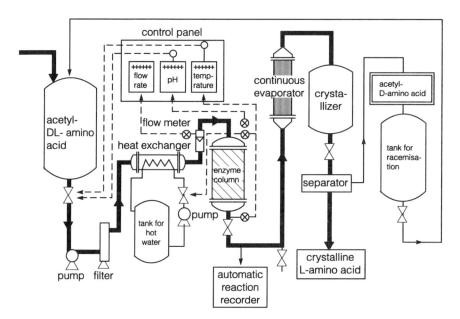

Figure 6.1. Flow diagram for the continuous production of L-amino acids by immobilized aminoacylase. (From Chibata, 1978, p. 172; reproduced by kind permission of Kadansha Scientific Ltd.)

for antibiotic synthesis provides further examples. Phenylglycine (whose D-enantiomer is a moiety of the semi-synthetic β-lactams ampicillin and cephalexin) is "classically" resolved as a diastereomeric salt, while its analogue, 4-hydroxyphenylglycine (which is a precursor of amoxicillin), is obtained by an enzymatic process. Indeed, there is a variety of published methods for this particular resolution, including the hydrolysis of the corresponding hydantoin, which yields N-carbamoyl-4-hydroxy-phenylglycine (28). The remaining hydantoin racemises at the alkaline pH of the reaction mixture, while the N-carbamoyl derivative does not. There is therefore a quantitative conversion of the initial racemate, and the product is easily converted to the corresponding amino acid using nitrous acid (Scheme 6.8).

The hydantoinases are particularly useful enzymes because the hydantoins are intermediates in the synthesis of many amino acids. Enzymes with specificities for the D- or the L-enantiomers are available. The enzyme from *Pseudomonas striata* is most active against dihydrouracil, and it is possible that this is its natural substrate. The most active substrates amongst the hydantoins are those formed from the neutral aliphatic amino acids, but several phenylglycine analogues are also hydrolysed.

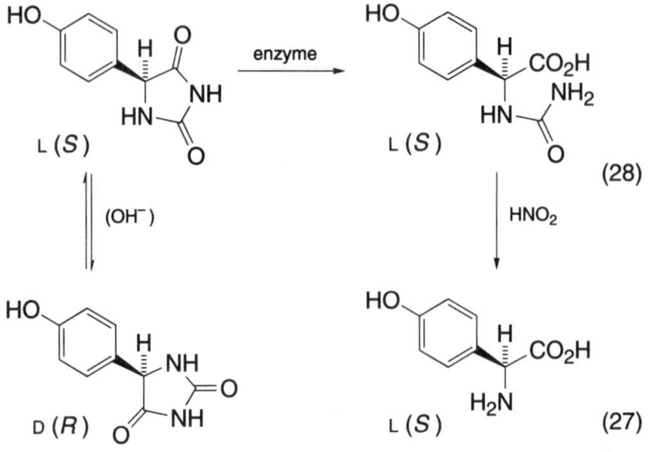

Scheme 6.8.

There are other published resolutions of 4-hydroxyphenylglycine, one of which involves the enzyme-catalysed hydrolysis of the ethyl ester (Scheme 6.9). Since the substrate is fully blocked at the amino and the carboxylate functions, it is scarcely soluble in water and must be dissolved in an organic solvent. The specific hydrolysis of the ester catalysed by an enzyme in an aqueous phase in contact with the organic solvent will yield a water-soluble carboxylic acid which transfers to the aqueous phase containing the enzyme. The conditions of the reaction therefore effect both hydrolysis of the ester and facile separation of the product. In fact, the enzyme is immobilized on a hydrophilic membrane at the interface between the two immiscible phases. The membrane reactor (Figure 6.2) comprises a large bundle of hollow fibres, each having an external diameter of about

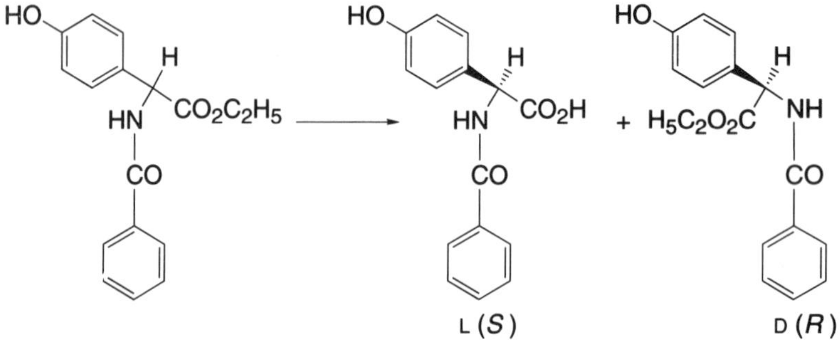

Scheme 6.9.

300 μm and a wall thickness of some 50 μm. The aqueous phase passes through the internal lumina of the fibres, while the organic phase flows over their external surfaces. The hydrolysis occurs as the ester passes across the membrane. This process, which Sepracor Inc. developed, may not be a fully commercial process for the synthesis of 4-hydroxy-D-phenylglycine, but it certainly represents an innovative use of an enzyme in a reaction where its substrate is poorly soluble in water. On the other hand, the Kanegafuchi company does use the hydantoinase process to manufacture some of the 1,200 tonnes of "amino acid" side chain needed to manufacture the estimated 2,800 tonnes of amoxicillin produced each year.

6.4.3 L-cysteine (29)

Much of the world's cysteine is isolated from keratin, mostly obtained in the form of hair. The process, which is based on an acid hydrolysis of the protein, is smelly, and it produces waste products which are difficult to treat. It is progressively being replaced by an enzymatic synthesis which once again couples a stereospecific hydrolysis with a racemisation process. Methyl-2-chloroacrylate is converted into DL-amino-Δ_2-thiazoline-4-carboxylate, and both enantiomers of this intermediate are then hydrolysed by an enzyme from *Pseudomonas thiazolinophilum* or *Sarcina lutea* directly to L-cysteine. The necessary racemisation occurs spontaneously during the reaction (Scheme 6.10).

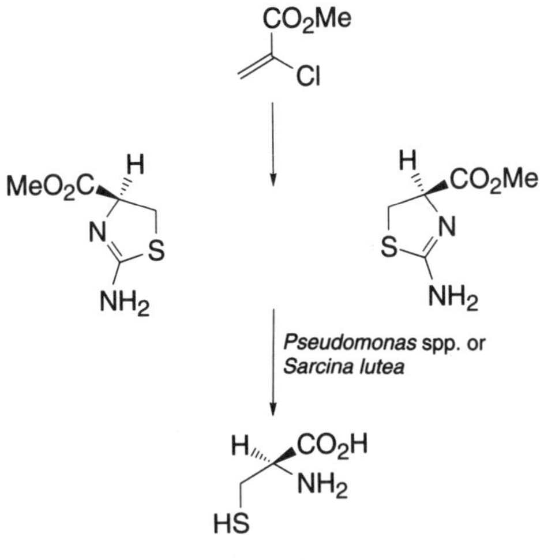

Scheme 6.10.

6.4.4 L-*aspartate (30)*

The outputs of L-phenylalanine (31) and of L-aspartate (30) have recently risen to supply the demand for the starting materials for the synthesis of

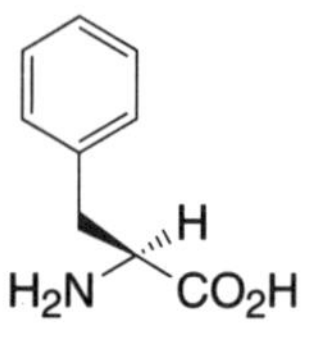

(31)

phenylalanine

the low-calorie sweetener Aspartame (8). The former amino acid is a fermentation product, while the latter is manufactured from fumaric acid and ammonia in an enzyme-catalysed process. The reacton is the reverse of one first discovered in 1926, in which resting cells of *E. coli* deaminate L-aspartate to fumarate. In the 1950s it was shown that growing cells of *E. coli* would accumulate L-aspartate when their growth medium was fed fumarate and ammonium ions. The enzyme responsible for this reversible reaction is L-aspartate ammonia lyase often called fumarase (Scheme 6.11).

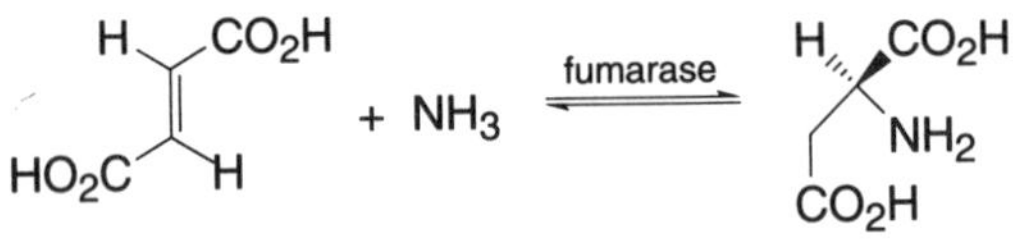

Scheme 6.11.

This stereoselective synthesis of the required enantiomer from a simple achiral precursor such as an alkene would be a preferred route for any chiral product. Even if a racemisation, such as that which accompanies the synthesis of L-methionine and L-cysteine, is practical, there are likely to be fewer steps in a selective synthesis from a carefully chosen achiral compound. In this instance the manufacture is very efficient. It is catalysed either by whole *E. coli* cells or by an enzyme preparation extracted from them. In

REACTOR CONFIGURATION

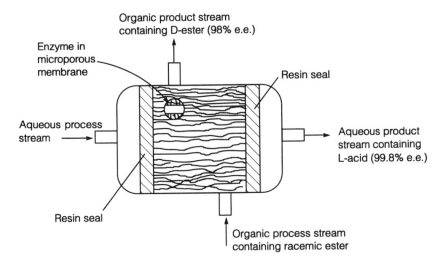

MEMBRANE CONFIGURATION

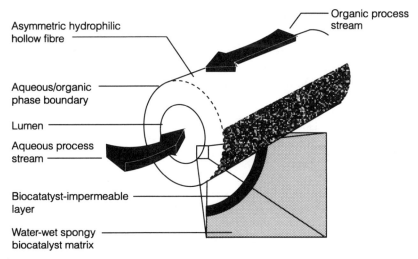

Figure 6.2. Schematic drawing of a two-phase membrane reactor. The membrane reactor consists of a bundle of thin hollow fibres. The ends of the fibre bundle are sealed with resin so that the aqueous process stream can enter the lumina of the fibres without mixing with the organic stream which flows over their outer surfaces. The enzyme is immobilized in the spongy matrix of the fibres. As the substrate diffuses from the organic phase into this matrix, it is hydrolysed. The product is soluble in the aqueous process stream which carries it out of the reactor. (Reproduced by kind permission of Sepracor Ltd.)

either case the catalyst is immobilized, for example in beads of a poly-acrylamide gel which are packed into a column. The reagents percolate through the column, as in the process for the manufacture of L-methionine (Figure 6.1). The L-aspartic acid formed is recovered from the effluent when acidified to pH 2.8. The yield of product is over 90%.

6.5 β-lactam antibiotics

Although the natural penicillins, G (32) and V (33), are useful antibiotics in their own right, their development as major pharmaceutical products was limited until a convenient method for the exchange of their 6-amido side chain was discovered. The new side chains increase the antibiotic activity of the β-lactam and improve their resistance to the bacterial β-lactamases, which will hydrolyse the β-lactam ring. These two effects create some very valuable commercial products whose annual sales exceed 10^9.

The β-lactam nucleus is very unstable below pH 2 and above pH 9, which precludes a simple acid- or base-catalysed hydrolysis to liberate the β-lactam nucleus 6-APA (6-aminopenicillanic acid) (34). Although this compound was known to be a minor component of the commercial *Penicillium acremonium* fermentation for the natural penicillins themselves, this also was not a suitable method for its production. A major breakthrough came around 1960 when several groups noticed that various micro-organisms were able to hydrolyse the *exo*-cyclic amide bond specifically, without also destroying the amide in the β-lactam ring (Scheme 6.12). The amidohy-

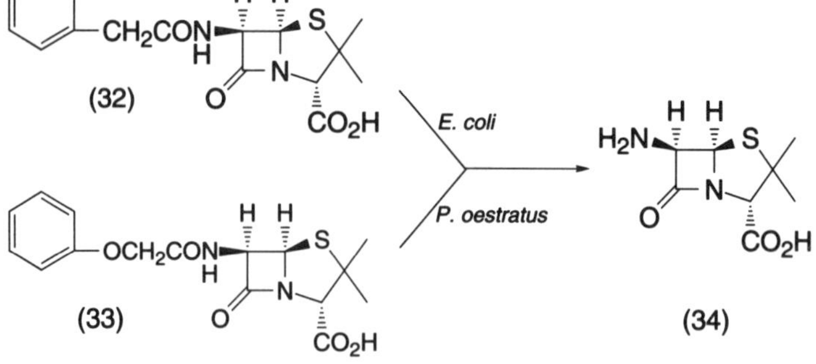

Scheme 6.12.

drolases responsible were specific either for penicillin G or for penicillin V; there was some hydrolysis of the less preferred substrate, but not enough to characterize one enzyme as being active against both substrates. These enzymes are frequently described in the literature as penicillin acylases.

The enzyme derived from *E. coli* will hydrolyse penicillin G to 6-APA, and this quickly became the basis of a manufacturing process. Some 10 years later a similar process for hydrolysis of penicillin V was developed, relying on an enzyme from a basidiomycete, *Pleurotus oestratus*. The reactions are usually performed at about pH 8, and the yield of recovered 6-APA is about 90%.

A chemical process was subsequently developed in which PCl_5 was used to break the amide bond. Partly for patent reasons, and partly because of the process chemist's general familiarity with chemistry rather than with enzymology, this chemical route competed effectively with the enzymatic catalysis for a number of years. This was particularly true for the hydrolysis of penicillin V, where for some time a suitable enzyme was not available. However, virtually all of the 16,000 tonnes of penicillin which is now converted to 6-APA is subject to enzymatic hydrolysis.

Although at high substrate concentrations (above about $50 \, g \cdot l^{-1}$) the reaction equilibrium of the enzymatic hydrolysis at pH 8 shifts noticeably away from complete hydrolysis, and at pH 5 lies towards amide synthesis, the reversal of the hydrolysis is not a useful method for resynthesizing the amide. The commercial value of 6-APA is about \$75 per kilogram, and some of the new side chains are equally valuable. Unless the amide synthesis is achieved in high yield, calculated from both the carboxyl and the amino halves of the new penicillin, the process is unlikely to be viable. High yields of this kind have not been recorded for the reversal of the enzymatic process, and until they are, chemical processes will remain the standard methods for adding the new side chains.

Cephalosporin C (35) is another important natural *β*-lactam antibiotic, isolated from fermentations involving *Acremonium chrysogenum*. It is related to penicillin G, and its synthesis is derived from the same metabolic intermediate, isopenicillin N (36) (Scheme 6.13). Unlike the penicillins, there is no therapeutic use for the natural product itself, and the manufacture of all useful cephalosporins requires the removal of the 7-aminoadipyl side chain to provide the 7-ACA (7-aminocephalosporanic acid) (37). This remains a chemical process in which NOCl or PCl_5 is used to break the *exo*-cyclic amide bond.

It is a curious fact that no single enzyme has shown more than a low activity in hydrolysing this particular amide link in cephalosporin C. However, the enzymatic preparation of 7-ACA is technically feasible, but an effective process requires two steps. If the 7-aminoadipyl function is first

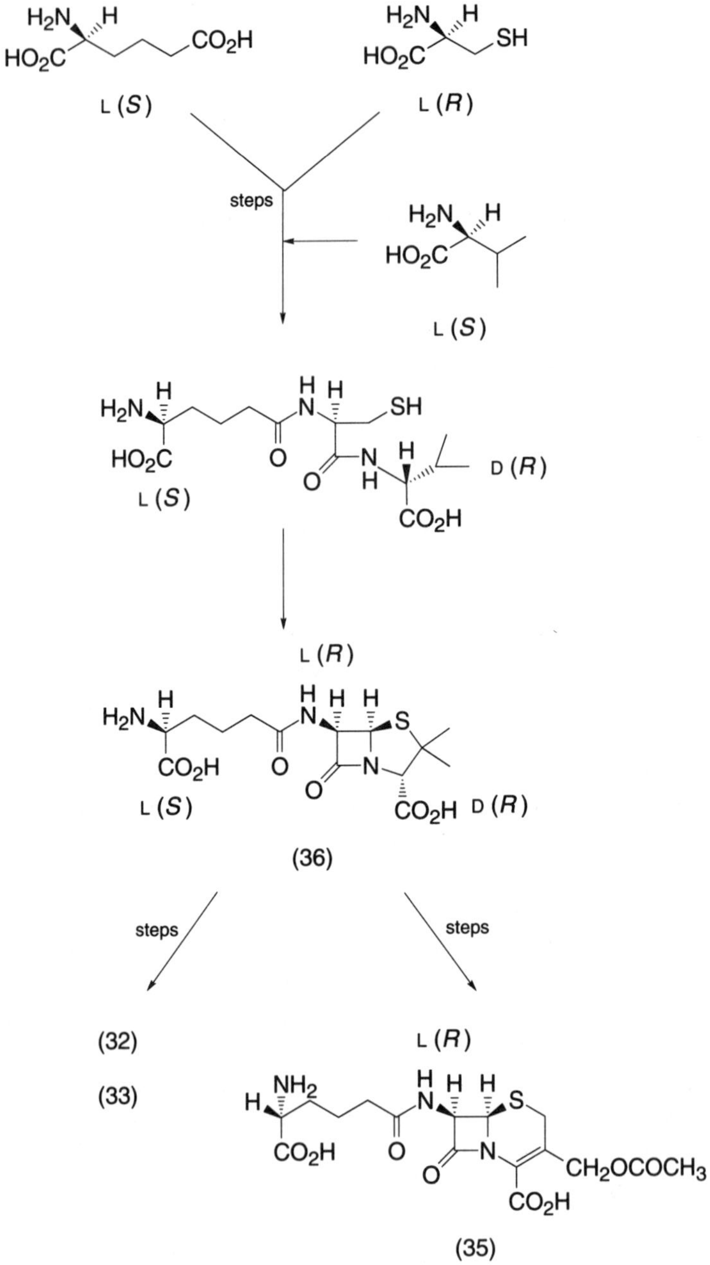

Scheme 6.13.

oxidized to a 7-glutaryl moiety, which is a reaction that the D-amino acid oxidase in *Fusarium solani* will catalyse, then an enzyme in *Pseudomonas diminuta* will hydrolyse this amide to yield 7-ACA (Scheme 6.14). However, this two-stage enzymatic process cannot compete with the chemical route.

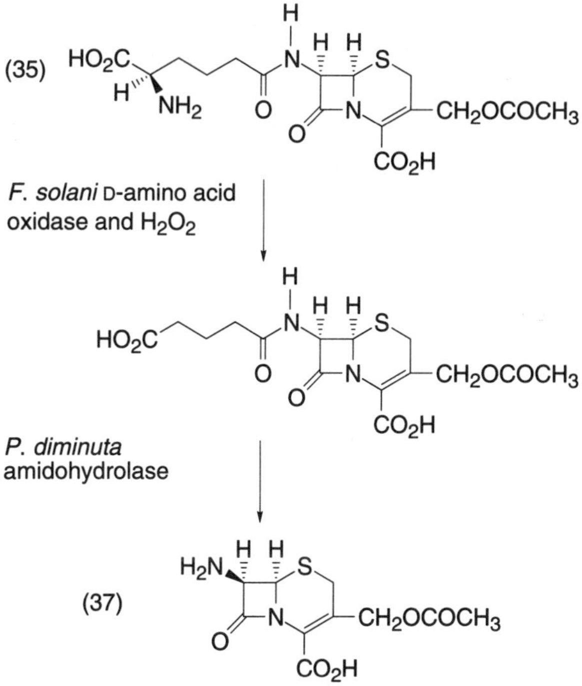

F. solani D-amino acid oxidase and H_2O_2

P. diminuta amidohydrolase

Scheme 6.14.

Although enzymes are of no commercial value in removing the amido side chain of cephalosporin C, they are useful in hydrolysing the 3-acetoxymethyl group (see Chapter 3, Scheme 3.6). This reaction uses an esterase from *Rhodosporidium toruloides*. It allows the synthesis of a range of modified cephalosporins, such as cefuroxime (38).

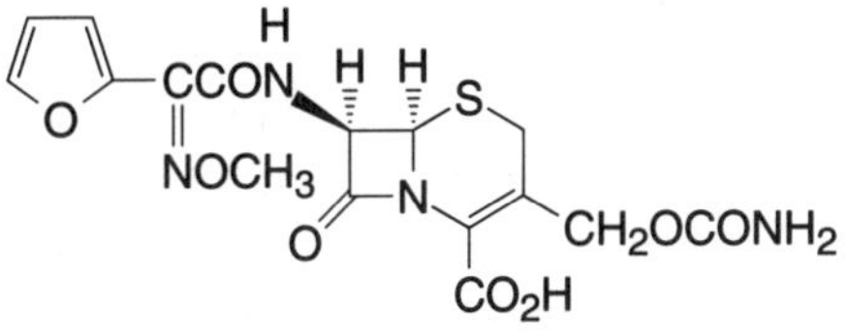

cefuroxime

6.6 Isoglucose

Enzymes, as distinct from fermentations, are rarely used in the large-scale processes whose products would rate as low-value commodity chemicals. There are, of course, many enzymatic steps in the manufacture of some foods and beverages, but these usually do not yield a well-defined chemical product. One exception is the manufacture of isoglucose, and the scale of this process makes it impossible to exclude from any description of the industrial use of enzymes.

Isoglucose is an equilibrium mixture of glucose and fructose prepared from cornstarch. It is a substitute for invert sugar, which is a 1:1 mixture of glucose and fructose prepared from sucrose. Starch is a mixture of the glucose polymers amylose (39) and amylopectin (40). The former contains

amylose

(39)

only 1–4α linkages in a linear structure; the latter is a highly branched structure built from a mixture of 1–4α and 1–6α links. The first steps in the manufacture of isoglucose hydrolyse the starch to glucose. This is practical only if the starch grains are first swollen and ruptured to form a gel, which occurs irreversibly at temperatures above 65 °C. The gels are very viscous, and at the high concentrations required for an economic hydrolysis of the starch, they would be impossible to handle if they were allowed to cool.

To keep the gel as mobile as possible, the suspension of starch grains is mixed with a heat-stable amylase and heated to 140 °C in a special jet cooker. At this temperature, the amylase, which catalyses the hydrolysis of the internal 1–4α links in the starch, is active for only about 30 seconds, but this is long enough to reduce the molecular weight of the starch and to thin the gel sufficiently that it can be handled at 100 °C. More amylase is then added, and after some 30 minutes of further hydrolysis the partially

amylopectin

hydrolysed starch can be cooled to about 55 °C, at which temperature it now remains fluid. Two other enzymes, amyloglucosidase and pullulanase, are then added which will catalyse the hydrolysis of both the 1–4α and the 1–6α links. The mixture of all three enzymes ensures that the bulk of the starch is hydrolysed to glucose, with less than 10% remaining as low-molecular-weight oligosaccharides (dextrins) (Scheme 6.15).

The emerging glucose syrup has a wide variety of uses. In the United States, large quantities, equivalent to some 4 million tonnes of glucose, are fermented each year to ethanol. A further 6 million tonnes (7.8 million tonnes world-wide) are treated with glucose isomerase. This enzyme inter-

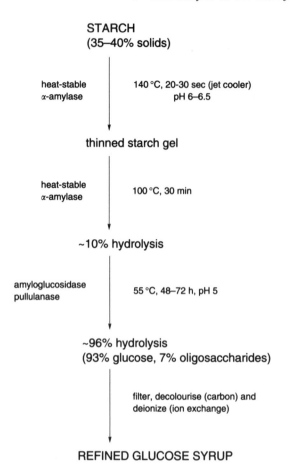

STARCH
(35–40% solids)

heat-stable
α-amylase

140 °C, 20-30 sec (jet cooler)
pH 6–6.5

thinned starch gel

heat-stable
α-amylase

100 °C, 30 min

~10% hydrolysis

amyloglucosidase
pullulanase

55 °C, 48–72 h, pH 5

~96% hydrolysis
(93% glucose, 7% oligosaccharides)

filter, decolourise (carbon) and
deionize (ion exchange)

REFINED GLUCOSE SYRUP

Scheme 6.15.

converts pairs of aldoses and ketoses, here glucose and fructose, creating
an equilibrium mixture of the two sugars (Scheme 6.16). This contains
about 51% glucose and 42% fructose, the remainder comprising oligosaccharides (high-fructose corn syrup); however, further large-scale chromato-
graphic processes can provide a cut containing 90% fructose (high-fructose
corn syrup) and 10% glucose. The raffinate, which is enriched in glucose
and contains some 10% oligosaccharides, is recycled firstly through a bed
of immobilized amyloglucosidase, which reduces the oligosaccharide content, and then, after mixing with the stream of fresh glucose syrup, through
the isomerisation process (Scheme 6.17).

The process is remarkable for its scale, for the range of intermediate
products (such as partially hydolysed starch, glucose syrup, and purified

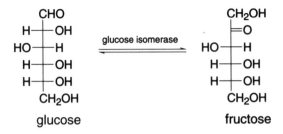

Scheme 6.16.

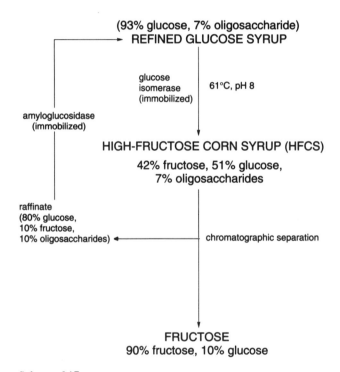

Scheme 6.17.

glucose) and for the nature of the enzymology. It provides a good example of the use of enzymes at high temperatures, as well as the large-scale use of immobilized enzymes. Although the early stages of the process use soluble enzymes, both the glucose isomerase and the amyloglucosidase are used as immobilized preparations. Finally, it is worth recalling the work of many of the nineteenth-century technologists whose ideas are now embodied in this process (see Chapter 1).

6.7 Synthesis of (*S*)-2-chloropropionate and acrylamide: the influence of cleaner chemistry

Biological catalysis is often considered to be an intrinsically clean technology. This is a dubious assertion, given the hazards and wastes associated with some fermentation processes. In fact, many developments in chemistry have offered improvements to existing processes and have been able to reduce the environmental burden of modern technology. Biological catalysis is no exception, and it does offer some cleaner chemistry. One major advantage is the possibility of keeping all fermentation and enzyme manufacture on one site, while allowing the use of the enzymes to be widely dispersed. If the enzymatic processes are simple, they should require fewer hazardous reagents than the chemical processes which they might replace. The dispersal of hazardous chemicals is thus limited, while the hazards accompanying the fermentation are centralized at one site. This is one advantage of the enzymatic process for the hydrolysis of the penicillins, which should limit the widespread use of PCl_5 at the expense of a few centralized *E. coli* fermentations.

The resolution of 2-chloropropionate and the hydrolysis of acrylonitrile are two examples of the use of new, cleaner technology to set alongside the new process for the manufacture of cysteine (see Section 6.4.3).

6.7.1 (S)-2-chloropropionate

The herbicides based on 2-aryloxypropionate are synthesized from 2-chloropropionate. Although only their *R*-enantiomers are active, they were originally marketed as racemates because of the added cost of resolving the mixture. The need to reduce the environmental burden of unnecessary chemicals has now placed a premium on a product containing only the active isomer. Moreover, if the resolution is performed at an early enough stage in the synthesis, there is a large cost to be saved in not wasting materials in the synthesis of the inactive isomer. This is important where the aromatic portion of the molecule is complex, as it is in fluazifop (41).

(41)

fluazifop

The optically active herbicides are synthesized from (S)-2-chloropropionate, since in this strategy the chiral centre is inverted when the aromatic portion is added. ICI started to address this problem by isolating a number of micro-organisms which were able to grow on chloropropionate. Amongst these were several *Pseudomonas* species which contained different enzymes for hydrolysing the R- and the S-enantiomers of the chloro acid. The product of the hydrolysis is lactic acid with an inverted chirality (Scheme 6.18). Since the (S)-2-chloropropionate is the required enantiomer, one strain of the micro-organism was genetically manipulated first to remove the enzyme which specifically hydolysed this enantiomer and second to enhance the yield of the enzyme with the opposite specificity. The enzyme in this organism is now the basis for resolving about 2,000 tonnes of (R,S)-chloropropionate each year.

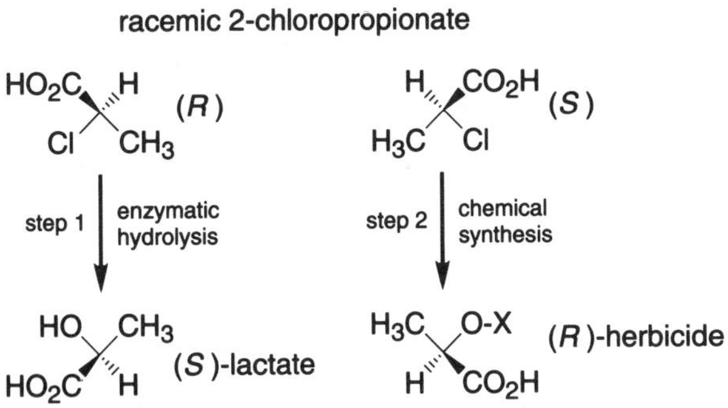

racemic 2-chloropropionate

HO_2C H   (*R*)   H CO_2H (*S*)
Cl CH_3 H_3C Cl

step 1 | enzymatic hydrolysis   step 2 | chemical synthesis

HO CH_3 H_3C O-X   (*R*)-herbicide
HO_2C H   (*S*)-lactate   H CO_2H

Scheme 6.18.

6.7.2 Acrylamide (42)

Micro-organisms contain a variety of enzymes which hydrolyse organic nitriles. These enzymes are often distinguished by their ability to attack aliphatic and aromatic substrates (Scheme 6.19). The Nitto Chemical

$$H_2O + CH_2{=}CH{-}CN \xrightarrow[\text{hydratase}]{\text{nitrile}} CH_2{=}CH{-}CONH_2$$
$$(42)$$

$$2H_2O + Ar{-}CN \xrightarrow{\text{nitrilase}} ArCO_2H + NH_3$$

Scheme 6.19.

Company in Japan now uses a nitrile hydratase from *Rhodococcus rhodochrous* to hydrolyse acrylonitrile to acrylamide. The established chemical process for this hydrolysis involves copper catalysis at 100 °C. It is difficult not only to prepare and recover the catalyst but also to separate and purify the acrylamide. The enzymatic process operates at 10 °C, so requires less energy; under such conditions there is less chance of either the acrylonitrile or the acrylamide polymerizing during the reaction. The recovered product is also of greater purity than the material from the chemical process.

This purer acrylamide can be polymerized to a higher molecular weight than is possible with the chemical product. Since much of the polyacylamide is used as a flocculant whose effectiveness increases with its molecular weight, smaller amounts of polymer prepared from the enzymatic acrylamide are needed for those precipitation processes in which it is the preferred flocculant. Some of these processes are found in the water-treatment industry, where a reduced input of a purer reagent is preferred. It therefore seems likely that the cleaner enzymatic process, which currently produces some 30,000 tonnes of acrylamide annually, will take a progressively larger share of the current annual production of about 200,000 tonnes.

6.8 The chiral switch

For small-scale chemistry, the practical value of the specificity and selectivity inherent in enzymatic catalysis is increasingly recognized. The topic is the subject of a number of recent books (see the Bibliography at the end of each chapter), and the current literature is full of examples. There may be some resistance to the large-scale use of enzymes, but there are now enough manufacturing processes in operation to demonstrate the effectiveness of the technology.

In addition to the commercial pressures outlined in the discussion of the trend to introduce cleaner chemistry (see Section 6.7), two strands of external regulation are likely to drive chemistry towards an increasing use of selective catalysis, whether derived from enzymes or from the many reactions in organic chemistry where stereochemical relationships are transferred from one chiral centre to another. The first regulator is the need for selective chemical processes which generate little waste and few side products. The second is the perception that an unwanted enantiomer in any product, and particularly a pharmaceutical one, is actually an impurity. At best it may be inactive, but at worst it can have a pharmacological activity in its own right which is radically different from the activity of the product itself. Thalidomide fell into the latter category, with disastrous results.

The authorities who licence pharmaceuticals have begun to argue that where the products are chiral, then the isomers should be separated and each rigorously tested on its own. Moreover, they believe it undesirable that an active chiral drug should be sold as a mixture with its inactive enantiomer, and companies will have to justify their decision to do so. The debate has already affected programmes of chemical synthesis in the pharmaceutical industry, with some groups resolving to synthesize only achiral compounds. Others will concentrate on the need for synthetic methods which are targeted at only one enantiomer. The drive towards chiral purity has emphasized the need for more effective methods of asymmetric synthesis, so that the separation of enantiomers is avoided. This change of attitude has become known as the "chiral switch".

Some of the fine-chemical manufacturers who supply the pharmaceutical industry now specialize in the use of enzymes for the synthesis of chiral synthons. Chiral cyanohydrins, as well as α- and β-amino and hydroxy acids, are just a few of the products now available, often produced directly from achiral precursors. Many of the hydroxy acids, such as citric acid (5) and both enantiomers of lactic acid, are fermentation products. The synthesis of some others (Scheme 6.20) resembles the manufacture of aspartate (see

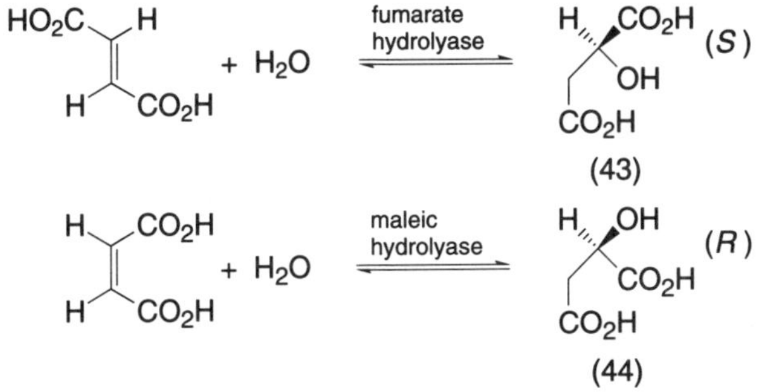

Scheme 6.20.

Section 6.4.4), both in the nature of the reaction [which requires a lyase to add water (rather than ammonia) across a carbon–carbon double bond] and in the use of growing cells. Purified fumarate hydrolyase is used to manufacture some 500 tonnes of L-(S)-malate (43) each year, while maleic hydrolyases will catalyse the synthesis of D-(R)-malate (44), a product for which there is as yet no substantial market. Such reactions are also important in the synthesis of both L-carnitine (45) and captopril (46) (Scheme 6.21).

The lactate dehydrogenases which are responsible in whole organisms for the synthesis of one or other enantiomer of lactic acid will also synthesize a range of substituted α-hydroxy acids from α-keto acids. The purified enzymes are used, and the reactions are more complex than those which the lyases catalyse, in that they require NADH as a reducing agent. This must

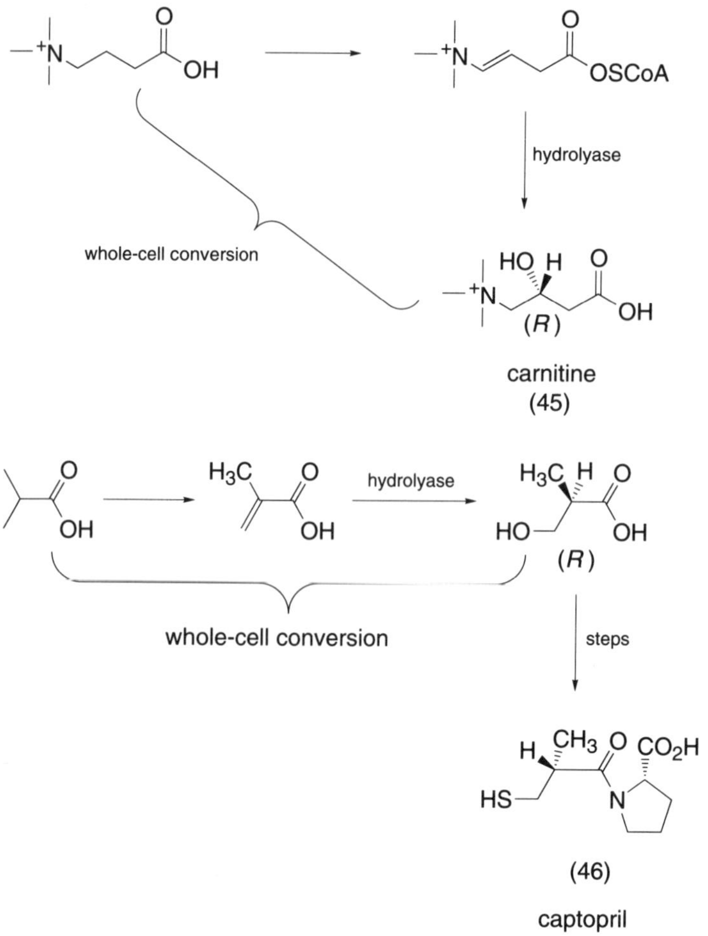

Scheme 6.21.

be recycled enzymatically, for example with an alcohol dehydrogenase (Scheme 6.22) (cf. Figure 4.5). The ultimate source of reducing energy may then be a chemical reduction which will prevent an increase in the concentration of oxidized co-product (acetaldehyde in Scheme 6.22) from affecting the equilibrium between the keto and hydroxy acids. Alternatively, formate dehydrogenase could be used, since the final oxidation products are CO_2 and water (Scheme 6.22).

D-glycerol-3-phosphate (*sn*-glycerol-3-phosphate) (47), which is used in phospholipid manufacture, is another synthon whose production requires the recycling of a co-factor. It is synthesized from glycerol and ATP, with

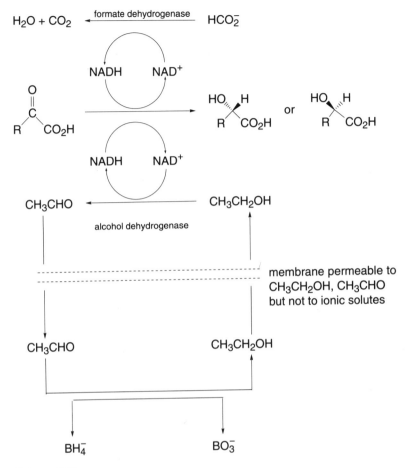

Scheme 6.22.

the enzyme glycerol kinase as catalyst, the ADP being recycled with a separate enzyme. The usual source of the necessary phosphate ester is phosphoenolpyruvate (Scheme 6.23), but polyphosphate and carbamoyl-phosphate are alternatives which deserve more attention. All of these are easily synthesized.

The suppliers of fine chemicals initially produce many new chiral compounds in kilogram amounts on the assumption that scale-up to an annual production of about 1 tonne is practical. The pharmaceutical industry almost certainly produces other synthons on this scale from its own resources. The outcome is likely to be an increasing use of enzymes in

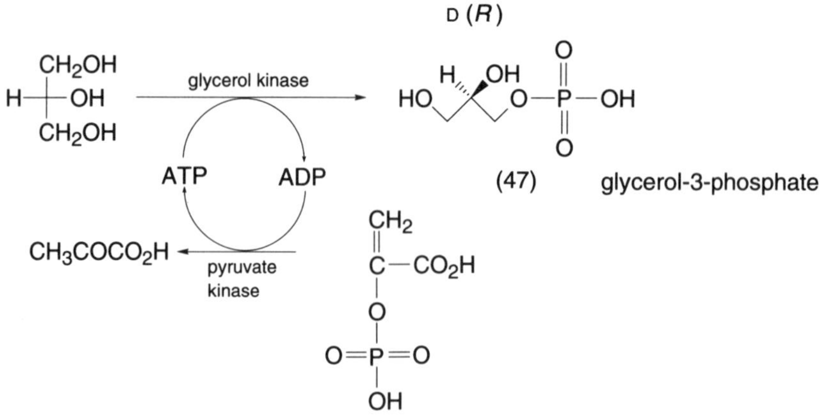

Scheme 6.23.

chemical manufacture. As they become more entrenched in syntheses at the laboratory and pilot scale, so will their use in manufacture increase. This is not to say that every process which uses an enzyme to make an effective product in the laboratory will have an enzyme employed in its manufacturing process. A development phase preceeds the latter stage, and this can radically alter the route of the synthesis. Nevertheless, particularly in the pharmaceutical industry, the constraints in bringing materials to the marketplace as rapidly as possible leave little time for the complete reworking of a process, so that it is likely that many initially incorporated enzymatic steps will remain. This is already presenting a challenge for the industry, which is finding some engineering problems in the scale-up of the technology.

6.9 Polymer synthesis

6.9.1 Nucleic acids

The directed synthesis of deoxyribonucleic acid (DNA) and its manipulation for genetic engineering would be impossible without enzyme catalysis. There is a range of restriction enzymes which can break DNA strands at specific points defined by the local base sequence, as well as ligases which can rejoin the broken ends where new base sequences are inserted, and they are essential catalysts. They complement the chemical synthesis of the oligonucleotides (small segments of DNA) which are inserted. A discussion of the technique is outside the scope of this chapter, largely because the scale of the synthesis is small. It is not a manufacturing process, but rather a facilitating technology of great significance. However, its value is evident from the influence of genetic engineering on modern biology.

6.9.2 Peptides

A combination of chemical and enzymatic catalyses is also used in the synthesis of proteins. The natural source of a protein is the organism which synthesizes it. If the supply is limited, then recombinant DNA technology allows the natural source to be replaced by a new one. The technique has allowed the animal proteins insulin and chymosin to be produced in large amounts in *E. coli* and other micro-organisms. Genetically engineered plant and animal cells, and indeed whole plants and animals, are also useful. Any one of these sources could provide tonne quantities of almost any protein as a highly purified product. Compared to the difficulty of purifying even small amounts of some proteins from their natural sources, this is a remarkable technical achievement. However, it is limited to the synthesis of proteins containing the 20 natural amino acids.

The enzymatic synthesis of peptides (Scheme 6.24) from which proteins can be constructed is not so limited, and chemical synthesis has an even wider application, but these are not yet suitable techniques for manufacture. Moreover, there are no general methods for building the peptides into full protein structures. Nevertheless, enzymes do have a role in the manufacture of peptides themselves. In a mixture of butan-1,4-diol and water, trypsin will catalyse the exchange of the carboxy-terminal alanine of porcine insulin with threonine *t*-butyl ester (Scheme 6.25). The reaction is essentially a transpeptidation in which the acyl group of lysine is transferred from one amino group on alanine to another on the threonine. This converts porcine insulin into the ester of the human hormone, and a simple deprotection yields one of the commercial products.

Thermolysin, which is another protease, will also catalyse peptide synthesis, and a new plant will shortly use this enzyme for the manufacture of the artificial sweetener Aspartame, at a scale of 2,000 tonnes per year. In this reaction the L-enantiomer of racemic phenylalanine methyl ester reacts specifically with the α-carboxyl group of *N*-protected L-aspartate (Scheme 6.26). Thus both the separation of the enantiomers of the phenylalanine and the protection of the γ-carboxyl group of the L-aspartate are unnecessary, which simplifies the synthesis. Although the equilibrium favours hydrolysis rather than synthesis, the peptide product, which is the *N*-protected precursor ester of Aspartame, forms an insoluble salt with the

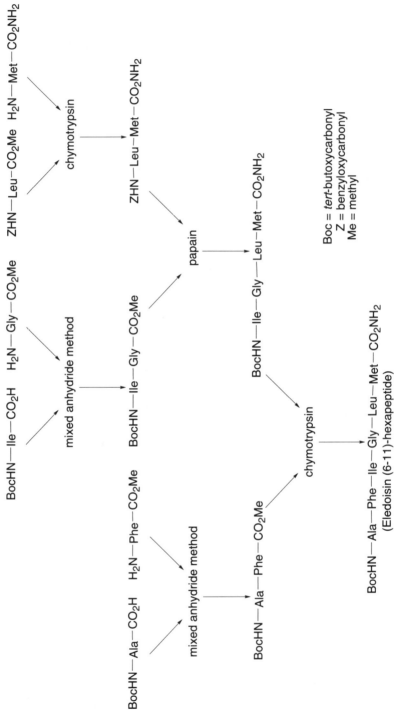

Scheme 6.24.

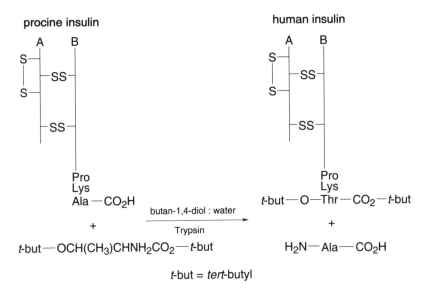

procine insulin
A
B
S
SS
S
SS
Pro
Lys
Ala — CO2H
+
t-but — OCH(CH3)CHNH2CO2 — t-but
butan-1,4-diol : water
Trypsin
human insulin
A
B
S
SS
S
SS
Pro
Lys
t-but — O — Thr — CO2 — t-but
+
H2N — Ala — CO2H
t-but = tert-butyl

Scheme 6.25.

H
CO2Me
NHZ
CO2H
H
CO2Me
NH2
racemise
H
NH2
CO2Me
thermolysin
recycle
H
CO2Me
NH
O
NHZ
H
+
CO2-
H
NH3+
CO2Me

Scheme 6.26.

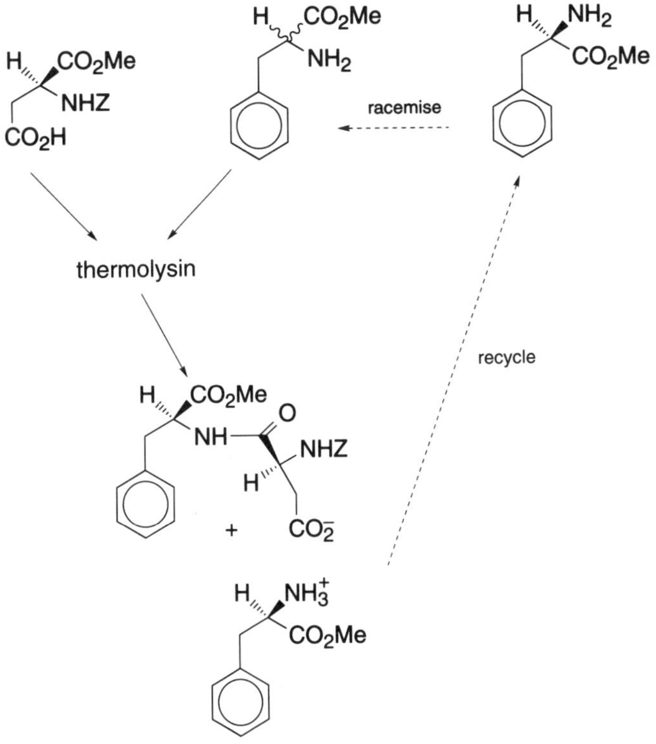

remaining D-phenyl-alanine methyl ester, and it precipitates. This pulls the reaction over in favour of synthesis, and the yield is almost quantitative. In a final chemical step, the *N*-protection on the aspartate moiety is removed chemically to form Aspartame. The residual D-phenylalanyl methyl ester is racemised and recycled through the process.

6.9.3 Oligosaccharides

The commercial manufacture and modification of oligosaccharides are important features of several industries. The processes are usually hydrolytic, but in a few instances enzymes catalyse a transglycosylation onto another carbohydrate rather than onto water, thus creating a new polymer. *β*-Cyclodextrin (48), which is a macrocyclic structure comprising seven molecules of glucose all linked 1-4α, is synthesized from starch with an enzyme from *Bacillus* species (Scheme 6.27). This cyclic carbohydrate acts

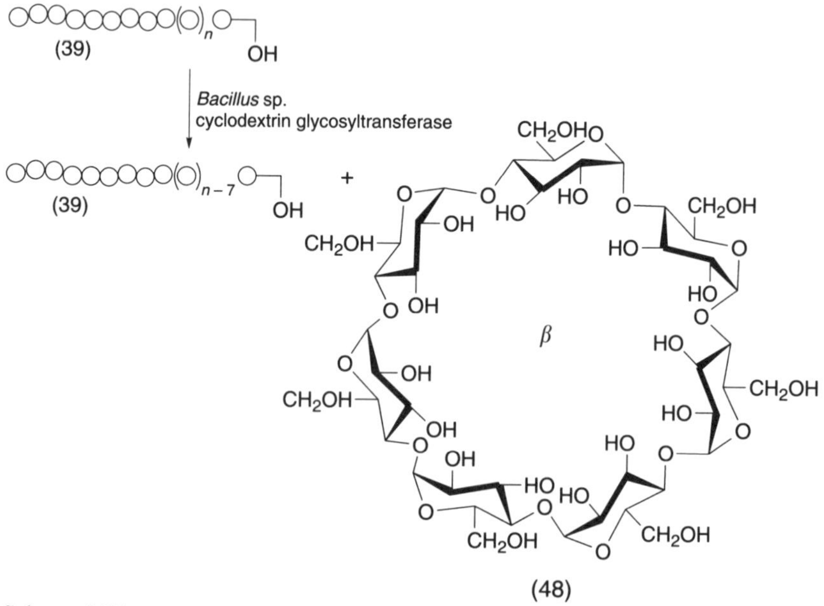

Scheme 6.27.

as a molecular host and will adsorb smaller molecules able to enter the centre of the macrocycle. In the food industry it is useful in capturing and retaining flavours, and it is also used in the formulation of pharmaceuticals. In 1989, about 850 tonnes were manufactured for a market which was rapidly increasing.

The more complex oligosaccharides attached to proteins have immuno-
logical and recognition functions which the pharmaceutical industry would
like to explore. They have structures comprising as many as 10 or 15
carbohydrate monomers in a branched structure which can be built up by
chemical procedures, but the numbers of protection and deprotection steps
necessary to ensure that the correct bonds are formed make the chemistry
very difficult. The synthetic strategies often make use of enzyme catalysis,
for which transglycosylation is the key reaction, either from simple donors,
as in the synthesis of cyclodextrin, or from the more complex nucleotide
substrates of the specific transferases (see Section 5.4). Over the next few
years this is likely to become a key target for larger-scale syntheses which
include enzymatic steps.

6.9.4 *Polyesters*

Some natural polyesters, such as polyhydroxybutyrate, are commercial
products of micro-organisms. Outside of the cell the enzymatic synthesis
of high-molecular-weight polyesters may be impractical on any sensible
scale, but the synthesis of smaller blocks with a molecular weight of several
thousand is practical. The ester bond is formed under conditions where
water activity is low, and the concentrations of alcohol and carboxylic acids
are high. The reaction (Scheme 6.28), which is a transesterification between

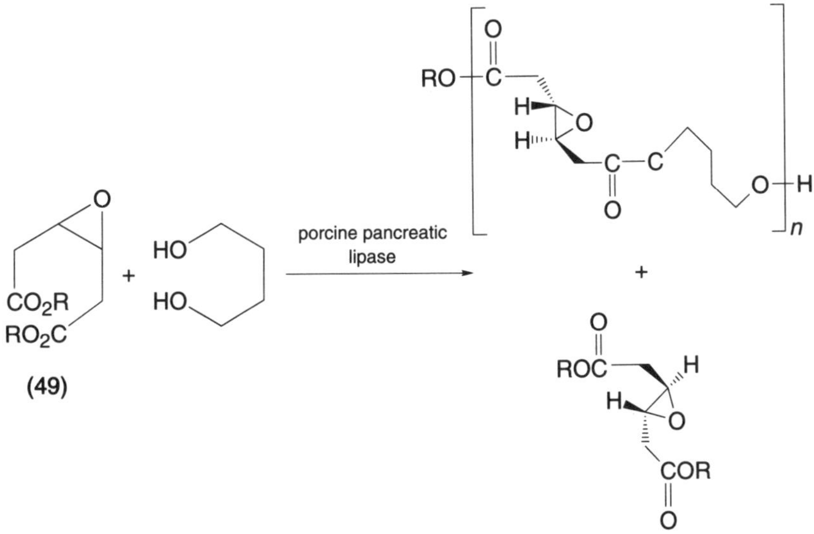

Scheme 6.28.

an activated ester of 3,4-epoxyadipate (49) and butan-1,4-diol, yields a chiral product which could prove a useful intermediate for the chemical synthesis of a chiral polymer. As with the synthesis of oligosaccharides, these are small-scale processes, but the manufacture and modification of speciality polymers could become an important target for enzyme chemistry.

6.10 New reaction conditions

The structure and stability of proteins are dependent on the presence of water, and enzymes are associated with catalysis in water. Indeed, this is one of their advantages, since they catalyse reactions over the range of temperatures at which water is liquid at normal atmospheric pressures. Nevertheless, there are exceptions, for example in the high-temperature digestion of starch (see Section 6.6). Nor does the aqueous environment in which enzymes usually act preclude their use as catalysts in organic solvents (see Sections 6.3, 6.4.2, 6.9.2, and 6.9.4).

There are good reasons for wanting to use enzymes outside their normal range. High temperatures may favour an increased solubility of the substrate, or at least a suitable colloidal state in the case of starch; they may prevent a process from becoming infected with invasive micro-organisms and thus are an aid to hygiene, if not sterility; finally, they may affect the equilibria of reactions in favour of the required products. Organic solvents are useful for the same reasons, but they may have an added advantage in partitioning out of the aqueous phase both substrates and products which can inhibit the catalysis.

However, proteins become less stable or unfold as temperatures increase, or when they are used in organic solvents. The chemistry of this process is sufficiently understood to allow it to be influenced by deliberate redesign of their polypeptide chains, by chemical modifications of their existing structures, or by careful choice of their reaction conditions. For example, the stability of proteins in organic solvent is dependent on the $\log P$ value of the solvent (Figure 6.3), so that the solvents with low $\log P$ values are considered less suitable for the catalysis than those with higher values. Moreover, some thermophilic micro-organisms, which naturally grow at high temperatures ($> 60\,^\circ\text{C}$), contain particularly stable enzymes which are useful as catalysts.

Often the enzymes are used in mixtures, where the water and the solvent are present as two distinct phases. This is useful where the substrate is poorly soluble in water. If the water content of the reactions involving hydrolytic enzymes is further reduced, then the solvent not only affects the

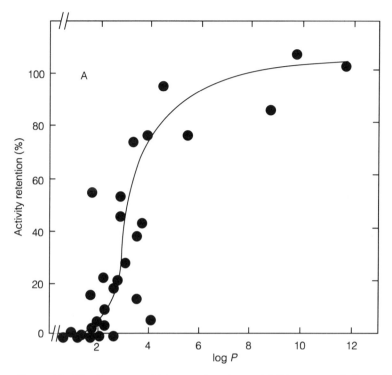

Figure 6.3. The degree to which organic solvents inhibit enzymes is dependent on their log *P* value. *P* is the partition coefficient of the solvent measured between octanol and water (i.e. *P* = solubility in octanol/solubility in water). Each point on the graph represents the activity of the enzyme in an reaction mixture containing water saturated with a solvent having the log *P* value shown. The reaction in this example is the microbial epoxidation of propene and butene. (From Laane et al., 1987, p. 73; reproduced by kind permission of Elsevier Science Publishers BV.)

overall concentrations of substrates and products at equilibrium but also can trap an unstable intermediate which might otherwise be hydrolysed. As Pottevin pointed out (see Chapter 1), the effect is to "reverse" the action of the hydrolytic enzyme, which now catalyses a synthetic reaction eliminating water from the substrates.

Already there are a few examples of large-scale reactions under such conditions (Table 6.6). Although it is likely that some are demonstration projects at a pilot scale, the transesterification of fats (see Scheme 6.29) is now a commercial reality in the manufacture of cocoa butter substitutes (50) and of methylethyltetradecanoate (isopropylmyristate) (51). The former is a valuable food ingredient, while the latter is used as a bulking agent in the formulation of cosmetics.

Table 6.6. *Large-scale processes catalysed by enzymes in organic solvents*

Catalyst	Reaction	Solvent
Nocardia corillina	Epoxidation	
	1-Octene	*n*-Hexadecane
	Styrene	*n*-Hexadecane
	1-Tetradecene	1-Tetradecene (substrate)
Horseradish peroxidase	Phenol polymerization	Ethylacetate
Trypsin (protease)	Transpeptidation	Butan-1,4-diol
Subtilisin (protease)	Ester hydrolysis	Dioxane, chloroform, etc.
Mucor miehei lipase	Transesterification	60–80 °C petroleum ether
Aspergillus niger	Ester synthesis	Substrate mixture
Arthrobacter simplex	Steroid dehydrogenation	Toluene

Source: Adapted from Lilly et al. (1990).

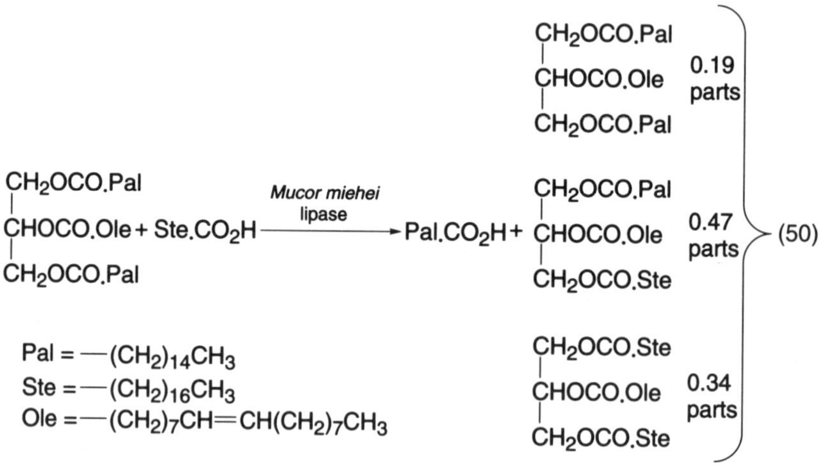

$$CH_2OCO.Pal$$
$$CHOCO.Ole + Ste.CO_2H \xrightarrow[\text{lipase}]{\textit{Mucor miehei}} Pal.CO_2H +$$
$$CH_2OCO.Pal$$

$$
\begin{array}{l}
CH_2OCO.Pal \\
CHOCO.Ole \quad \text{0.19 parts} \\
CH_2OCO.Pal
\end{array}
$$

$$
\begin{array}{l}
CH_2OCO.Pal \\
CHOCO.Ole \quad \text{0.47 parts} \\
CH_2OCO.Ste
\end{array}
\Bigg\} \; (50)
$$

$$
\begin{array}{l}
CH_2OCO.Ste \\
CHOCO.Ole \quad \text{0.34 parts} \\
CH_2OCO.Ste
\end{array}
$$

Pal = —$(CH_2)_{14}CH_3$
Ste = —$(CH_2)_{16}CH_3$
Ole = —$(CH_2)_7CH{=}CH(CH_2)_7CH_3$

Scheme 6.29.

$$
\begin{array}{l}
H_3C \\
{\diagdown}CHO.CO.(CH_2)_{12}CH_3 \\
H_3C {\diagup}
\end{array}
$$
(51)
isopropylmyristate

The control of water activity is a significant technical problem when the enzyme is in nearly dry solvent. Where a lipase catalyses ester synthesis, rather than transesterification, stoichiometric amounts of water are released.

Although some water is essential for enzyme activity, too much will affect the balance between ester synthesis and hydrolysis. Effective scale-up will require better methods of control than are presently available.

6.11 Genetic engineering

6.11.1 Protein production and protein engineering

The practice of genetic engineering can now transfer the synthesis of almost any protein into a micro-organism, or some other cell which will grow in a fermenter. Difficulties which once seemed as if they might affect the scope of the work must now be regarded as technical rather than fundamental limits. The commercial value of the recombinant proteins such as insulin, erythropoietin and interferon is considerable (Table 6.7), but in terms of sheer production the manufacture of chymosin is remarkable. The natural source of this enzyme, which is a specific protease used in cheese manufacture, is calf stomach. Its manufacture is now progressively being transferred to recombinant organisms, and each year Pfizer now markets over 2 tonnes of Chymax, which is manufactured using a recombinant *E. coli* strain.

The importance for biocatalysis is clear; however restricted are the amounts of a useful biocatalyst which can be obtained from its natural source, the transfer of its production to a recombinant organism could supply any reasonable demand. This powerful technique now raises other possibilities for the future development of biocatalysis.

The activity of an enzyme is dependent on its three-dimensional structure, and the forces which fold it into this shape are essentially delineated in the primary sequence of amino acids in the peptide chain. The specificity and selectivity of the catalysis result from the interaction of the substrate with

Table 6.7. *Estimated world-wide sales (1991) of recombinant proteins which are used as pharmaceuticals*

Recombinant product	Estimated sales ($M)
Erythropoietin	990
Human insulin	540
Human growth hormone	525
α-Interferon	445
Tissue plasminogen activator	270
Granulocyte colony stimulating factor	255

Source: Decision Resources Inc.

the side chains of a few nicely positioned amino acids in the protein. The analysis of these interactions usually relies on X-ray crystallography, either of the protein itself or of one closely related to it. Some rules are beginning to emerge about the way in which proteins fold. Their overall shape is often conserved amongst related enzymes, despite considerable changes in their primary amino acid sequence. This allows the gross features of the catalytic centre to be predicted without an X-ray structure, and the general nature of the catalysis often can be described when this structural information is combined with other data from the reaction kinetics and from amino acid substitutions which genetic engineering can introduce. Unfortunately, some important features of the catalysis seem to depend on the precise locations of interactions which are below the limit of resolution of X-rays.

Despite this rather nebulous state of affairs, some changes can be made which affect the enzyme in a predictable fashion. The exchange of amino acids which are easily oxidized (e.g. methionine residues) can improve the stability of the enzyme, but usually at the expense of lower catalytic activity. The incorporation of additional disulphide bridges will also increase stability, but usually not beyond what is available from the native enzyme under ideal conditions; the conditions have to be rigged to show the improvement. Changes in the amino acids around the active site can affect the catalytic properties in a useful fashion, but not beyond what can be achieved by chemical modification (see Section 6.9.2) or by random mutation and selection.

It is wrong to dismiss these efforts as impractical. They provide much valuable information about the chemical mechanism of enzyme catalysis. Once the basic rules are understood, they hold a promise for the rational redesign of enzymes.

6.11.2 Metabolic pathway engineering and ascorbic acid manufacture

Of greater immediate relevance for this industrial biocatalysis is the use of genetic engineering to incorporate new proteins into a cell. The commercial products of this technology, such as insulin and chymosin, do not affect the metabolism of the producing cell; they are end-products of their own right. If, on the other hand, the cell is altered to produce a new enzyme which is metabolically active in the cell itself, it becomes possible to extend a metabolic pathway by one or two desirable steps, an action which cannot be achieved through the normal process of mutation and selection, which can only act on the complement of enzymes which the cell already contains. This concept has been applied to a number of commercial processes, one of which is the manufacture of ascorbic acid.

Over the years there have been various attempts to simplify the Reichstein synthesis (see Chapter 1), which is now responsible for the manufacture of about 50,000 tonnes of ascorbic acid each year. The first attempts were mostly based on the microbial oxidation of glucose directly to 5-ketogluconate, with the ubiquitous *Acetobacter xylinum* discovered by Brown. Other organisms, including other *Acetobacter* sp. and *Erwinia* sp., will take the oxidation of glucose further, to 2,5-diketo-D-gluconate (52). Chemical reduction then produces a mixture of 2-keto-D-gluconate (53) and 2-keto-L-gluconate (54), depending on the selectivity of the reduction of the carbonyl group, but microbial reduction, which with *Brevibacterium* sp. or with *Corynebacterium* sp. is both regiospecific and stereoselective, yields only the latter, and in good yield (Scheme 6.30). This is the final inter-

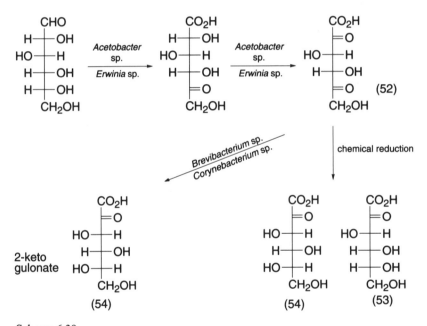

Scheme 6.30.

mediate in the route to ascorbate in the Reichstein synthesis. These two steps of oxidation and reduction can be run together in a mixed fermentation. *Acetobacter* and *Corynebacterium* mutants, used sequentially, will convert as much as 80% of the glucose to 2-keto-L-gulonate, and this is the limit of what could be achieved with the microbiology which predated the introduction of genetic engineering techniques.

The biotechnology company Genentech Inc. made use of these new techniques. The 2,5-diketo-D-gluconate reductase was purified from

Corynebacterium, and from the knowledge of the amino acid sequence which was gained, they probed for and then recovered the gene for this enzyme. They then transferred the gene into *Erwinia herbicola*, which expressed the reductase along with its own enzyme responsible for oxidizing glucose to 2,5-diketo-D-gluconate (52). This recombinant organism can synthesize 2-keto-L-gluconate (54) directly from glucose, achieving a radical simplification of the Reichstein process. In the process, the glucose provides ascorbate such that the orientation of the carbon skeleton is reversed compared to the orientation of its incorporation in the Reichstein synthesis (cf. Schemes 6.30 and 1.10), a nice demonstration of the power of biological catalysis. A final concentration of $20 \, \mathrm{g \cdot l^{-1}}$ is obtained after a 72-hour fermentation, with just under a 50% yield based on the input of glucose.

The further potential of the technology is seen in a recent manipulation of the synthesis of cephalosporin C (35). If the two enzymes (Section 6.5) which are needed to remove the 7-aminoadipyl side chain are both transferred to *A. chrysogenum*, then the engineered strain will produce 7-ACA (37) directly.

With such recombinant micro-organisms with new metabolic pathways to hand it is interesting to reflect on why neither these nor other examples are yet used in manufacture. Ultimately the answers are economic rather than technical. The high catalytic activity of enzymes is often offset by a low process intensity; few reactions are possible at the high reagent concentrations which the process chemist prefers. The oxidation of D-glucitol to L-sorbose is an exception, high yields being possible at concentrations above $250 \, \mathrm{g \cdot l^{-1}}$. Moreover, it is a well-integrated process. D-glucitol is an important product in its own right, with annual sales of about 0.6 million tonnes; the 70-m^3 fermentation is continuous; hydrogen for the reduction of glucose and chlorine for the oxidation of the diacetone adduct of L-sorbose are generated together by the electrolysis of sodium chloride; finally, the diacetone adduct of 2-keto-L-gulonate is easily recovered. In contrast, the direct microbial synthesis requires a low glucose concentration (about $40 \, \mathrm{g \cdot l^{-1}}$); 2-keto-L-gulonate itself is not easily recovered, nor is any D-glucitol produced.

Similar problems would affect the manufacture of 7-ACA (37) with a recombinant strain of *A. chrysogenum*, for which new conditions for the fermentation and recovery processes would have to be developed. It is often hard for new processes to displace existing ones; it is probably better to view the recent attempts to simplify these syntheses as examples of what modern strategies promise for entirely new processes. At some point a new product will arise out of these methods which would not have been available in any other way. It could well be soon, given the speed with which this technology is advancing. There should by now be enough familiarity with large-scale biocatalysis using enzymes and micro-organisms to allow the product to be manufactured successfully.

Bibliography

Andresen, O. (1993). Production of Bulk Chemicals With the Use of Enzymes. *Chimia, 47,* 58.

Chibata, I. (1978). *Immobilized Enzymes.* Halsted, New York.

Collins, A. N., Sheldrake, G. N., & Crosby, J. (1992). *Chirality in Industry: The Commercial Manufacture and Applications of Optically Active Compounds.* Wiley, Chichester.

Crueger, W., & Crueger, A. (1989). *Biotechnology: A Textbook of Industrial Microbiology,* 2nd ed. Sinauer Associates, Sunderland, Mass.

Gerhartz, W. (1990). *Enzymes in Industry: Production and Application.* VCH Publishers, Weinheim.

Kulla, H. G. (1991). Enzymatic Hydroxylations in Industrial Applications. *Chimia, 45,* 81.

Laane, C., Boeren, S., Hilhorst, R., & Veeger, C. (1987). *Biocatalysis in Organic Media.* Elsevier, Amsterdam.

Lilly, M. D., Dervarkos, G. A., & Woodley, J. M. (1990). Two-Phase Biocatalysis: Choice of Phase Ratio. In *Opportunities in Biotransformation,* ed. L. G. Copping et al., p. 13. Elsevier, Amsterdam.

Sheldon, R. (1990). Industrial Synthesis of Optically Active Compounds. *Chem. Ind.,* 212.

Wolnak, B., & Scher, M. (1990). *Industrial Use of Enzymes: Technical and Economic Barriers.* Bernard Wolnak & Associates, Chicago.

Index